Alpheus Spring Packard, Frederic Ward Putnam

The mammoth cave - descriptions of the fishes, insects and crustaceans found in the cave

including figures of the various species, and an account of their allied forms

Alpheus Spring Packard, Frederic Ward Putnam

The mammoth cave - descriptions of the fishes, insects and crustaceans found in the cave
including figures of the various species, and an account of their allied forms

ISBN/EAN: 9783742839367

Manufactured in Europe, USA, Canada, Australia, Japa

Cover: Foto ©berggeist007 / pixelio.de

Manufactured and distributed by brebook publishing software
(www.brebook.com)

Alpheus Spring Packard, Frederic Ward Putnam

The mammoth cave - descriptions of the fishes, insects and crustaceans found in the cave

THE

MAMMOTH CAVE

AND ITS

INHABITANTS,

OR DESCRIPTIONS OF THE

FISHES, INSECTS AND CRUSTACEANS

FOUND IN THE CAVE;

WITH FIGURES OF THE VARIOUS SPECIES, AND AN ACCOUNT OF ALLIED FORMS, COMPRISING NOTES UPON THEIR STRUCTURE, DEVELOPMENT AND HABITS, WITH REMARKS UPON SUBTERRANEAN LIFE IN GENERAL.

BY

A. S. PACKARD, JR., AND F. W. PUTNAM,

EDITORS OF THE AMERICAN NATURALIST.

SALEM:
NATURALISTS' AGENCY.
1872.

PRINTED AT THE

SALEM PRESS.

Corner of Liberty and Derby Streets,

SALEM, MASS.

F. W. PUTNAM & CO.

PREFACE.

THE following pages were first published in the "American Naturalist" for December, 1871 and January, 1872, with the exception of the Synopsis of the family including the Blind fishes of the cave, which was first published in the "Annual Report of the Peabody Academy of Science for 1871."

In bringing the several articles together in the present form but slight changes have been made, principally in the form of a few additional notes.

It will undoubtedly be the good fortune of some visitors to the cave to discover other kinds of animals than those mentioned in the following pages, and to observe new facts relating to the habits of the various species. For it must be remembered that all the observations thus far recorded have been made by but a very few of the thousands who annually visit the Mammoth Cave, and that no thorough zoological exploration of the cave has yet been undertaken. Should any new facts be observed, or unknown species discovered, the authors of this little work would be pleased to be informed of them, and communications on all such matters are solicited for publication in the pages of the AMERICAN NATURALIST.

THE AUTHORS.

PEABODY ACADEMY OF SCIENCE, *Salem, Mass., Feb.* 1872.

THE MAMMOTH CAVE

AND ITS

INHABITANTS.

CHAPTER I.

THE FORMATION OF THE CAVE.*

BY F. W. PUTNAM.

AFTER the adjournment of the meeting of the American Association for the Advancement of Science, held at Indianapolis, in August last, a large number of the members availed themselves of the generous invitation of the Louisville and Nashville Railroad Company, to visit this world renowned cave, and examine its peculiar formation and singular fauna.

The cave is in a hill of the subcarboniferous limestone formation in Edmondson County, a little to the west and south of the centre of Kentucky. Green river, which rises to the eastward in about the centre of the state, flows westward passing in close proximity to the cave, and receiving its waters thence flows northwesterly to the Ohio.

The limestone formation in which the cave exists, is a most interesting and important geological formation, corresponding to the mountain limestone of the European geologists, and of considerable geological importance in the determination of the western coal fields.

We quote the following account of this formation from Major S. S. Lyon's report in the fourth volume of the Kentucky Geological Survey, pages 509–10.

"The sinks and basins at the head of Sinking creek exhibit in a striking manner, the eroding effect of rains and frost — some of the sinks, which are from forty to one hundred and ninety feet

---

* From the AMERICAN NATURALIST for December, 1871.

deep, covering an area of from five acres to two thousand. The rim of sandstone surrounding these depressions is, generally, nearly level; the outcropping rocks within are also nearly horizontal. Near the centre there is an opening of from three to fifteen feet in diameter; into this opening the water which has fallen within the margin of the basin has been drained since the day when the rocks exposed within were raised above the drainage of the country, and thus, by the slow process of washing and weathering, the rocks, which once filled these cavities, have been worn and carried down into the subterranean drainage of the country. All this has evidently come to pass in the most quiet and regular manner. The size of the central opening is too small to admit extraordinary floods; nor is it possible, with the level margin around, to suppose that these cavities were worn by eddies in a current that swept the whole cavernous member of the subcarboniferous limestone of western Kentucky; but the opinion is probable that the upheaving force which raised these beds to their present level, at the same time ruptured and cracked the beds in certain lines; that afterwards the rains were swallowed into openings on these fractures, producing, by denudation, the basins of the sinkhole country, and further enlarging the original fractures by flowing through them, and thus forming a vast system of caverns, which surrounds the western coal field. The Mammoth Cave is, at present, the best known, and, therefore, the most remarkable."

So much has been written on the cave and its wonders, that to give a description of its interior would be superfluous in this connection, even could we do so without unintentionally giving too exaggerated statements which seems to be the natural result of a day underground, at least so far as this cave is concerned, for after reading any account of the cave, one is disappointed at finding the reality so unlike the picture. As the Association party was accompanied by one * who, while a most enthusiastic collector and explorer, was also a calm recorder of statements made by the geologists of the party, we cannot do better in conveying to our readers the general geological character and structure of the cave than to copy his account.

" As we expected to remain within the cave a long time, our trusty guide, Frank, had provided himself with a well-filled can of oil, to replenish our lamps, and with this strapped upon his back he led the way into the thick darkness. We shall attempt no description of the cave. Its darkness must be felt to be appreciated, and no form of expression, understood by mortals who have never descended to its cavernous depths, nor trod its gloomy

---

* W. P. FISHBACK, Esq., of the Indianapolis Daily Journal.

corridors, can convey anything like an adequate idea of the place. After spending fifteen hours within its chambers, it is absolutely nauseating to read the descriptions which have been current in the letters of newspaper correspondents for a quarter of a century, and even the vigorous and picturesque language of Bayard Taylor becomes tame and commonplace when it attempts to describe this subterranean wonder of the world.

How and when the cave was made, were the leading questions in the minds of the geologists. They do not believe that the cave was the immediate result of some violent upheaval of the strata, which left these vast crevices and chambers of which the cave is composed ; neither do they share the popular belief that the rapid and violent action of some subterranean stream of water has worn these deep channels through the limestone ; on the contrary, they find conclusive evidence that the same agencies are at work and the same changes in progress to-day that have been slowly, steadily and quietly, through vast periods of time, accomplishing the marvellous wonders that now astonish the beholder. The cave is wrought in the stratum known as the St. Louis limestone, which in some places reaches a thickness or depth of four hundred feet. This stone is dissolved whenever it is subjected to the influence of running or dripping water impregnated with carbonic acid gas. Water exposed to the air readily absorbs this gas, and surface water percolating through small fissures of the limestone, dissolves it. Another fact should be stated. When, during this process of solution, the water becomes thoroughly impregnated with lime, it loses its power to dissolve the stone. The following conditions, then, were essential to the productions of the cave, assuming what is not disputed by geologists, that the place where the cave now is, was once nearly solid limestone. First, that there should be fissures in the strata, allowing the ingress of the surface water. Secondly, there should be a place or places of exit for the water charged with limestone in solution. Without the latter, the water would become charged with lime, fill up the crevices, and the dissolving process would cease. These conditions are all present to-day, and have remained the same during the countless ages that have passed away while the work has been in progress. There have doubtless been times in the history of the cave, when, owing to a greater flow of water, the work has progressed more rapidly than at present, but that the results have been accomplished in the manner stated, rather than by the process of attrition by rapid currents of large volumes of water, seems to be the general opinion of scientific men. This theory is strengthened by the fact that where the cave attains its greatest heights, and reaches its lowest depths, the dripping waters have never ceased their labors, and are busily at work to-day. In the Mammoth Dome, for instance — rarely seen by visitors, on account of the dangers and fatigue incident to the journey — where the chasm attains a height and depth of more than two hundred and fifty feet, a cascade falls from a great height, and keeps the entire surface of the rocks covered with dripping water. This, falling into a deep pit below, finds an exit through which it bears away a portion of the lime composing the rock. After a walk of thirteen hours, our guide informed us that he would conduct us to the Mammoth Dome if we felt able to bear the fatigue of the journey. Foot-sore and weary, we were not in a favorable condition for so arduous an undertaking, but Mr. Thomas Kite of Cincinnati, who had visited the locality thirty years ago, urged us to go, and told us the sight of this Dome was worth all the rest. Provided with magnesium and calcium lights, we crawled and climbed our way to the brink of the pit, the bottom of which was reached by a rickety ladder, slippery and dripping with water. A portion of the party descended, and when all were ready the lights were ignited, and the immense dome was revealed to us in all its majestic beauty. Upon our return, three hearty cheers were given to the good friend at whose earnest solicitation we undertook this part of our journey.

We are indebted to Professor Alexander Winchell, of the University of Michigan, for the following abstract of his views concerning the formation of the cave.

'The country of the Mammoth Cave was probably dry land at the close of the coal period, and has remained such, with certain exceptions, through the Mesozoic and Cænozoic ages, and to the present. In Mesozoic times, fissures existed in the formation, and surface waters found their way through them, dissolving the limestone and continually enlarging the spaces. A cave of considerable dimensions probably existed during the prevalence of the continental glaciers over the northern hemisphere. On the dissolution of the glaciers, the flood of water which swept over the entire country, transporting the materials which constituted the modified drift, swept through the passages of the cave, enlarging them, and leaving deposited in the cave, some of the same quartzose pebbles which characterize the surface deposits from Lake Superior to the Gulf of Mexico. Since the subsidence of the waters of the Champlain epoch, the cave has probably undergone comparatively few changes. The well one hundred and ninety-eight feet deep, at the further end of the cave, shows where a considerable volume of the excavatory waters found exit. The Mammoth Dome indicates probably, both a place of exit and a place of entrance from above. So of the vertical passages in various other portions of the cave.'

We believe that the views of Professor Winchell are in harmony with those of the other eminent geologists of the party, and when it is considered that the geologists of this excursion stand in the front rank of the most eminent scientific men of the world, their views upon this interesting subject are well worthy of attention. Before dismissing this branch of the subject, we will take occasion to correct a popular error concerning the formation of the beautiful structures that adorn the ceilings of some portions of the cave. In the dryer localities, where the floors are dusty and everything indicates the prolonged absence of moisture, the ceiling is covered with a white efflorescence that displays itself in all manner of beautiful shapes. It requires no stretch of the imagination to discover among these, the perfect forms of many flowers. The lily form prevails, and the ceilings of many of the chambers are covered with this beautiful stucco work, surpassing in delicacy and purity the most beautiful workmanship of man. These are not produced, as many suppose, by the dripping of water, and the gradual deposit of sulphate of lime upon the outer portions. The stalactite is formed in this manner, but these are neither stalactiform, nor are they produced in a similar way. Dripping water is the agency that forms the stalactite, while the efflorescence in the dryer portions of the cave cannot take place where there is much moisture. The growth of these beautiful forms is from within, and the outer extremities are produced first. They are the result of a sweating process in the limestone that forces the delicate filaments of which they are composed through the pores upon the surface of the rock, their beautiful curved forms resulting from unequal pressure at the base, or friction in the apertures through which they are forced. Mr. L. S. Burbank, of Lowell, Mass., has kindly furnished us with the following abstract of his opinions upon this interesting subject.

'The rosettes, wreaths, and other curved fibrous forms of gypsum, in the Mammoth Cave, occur only in particular strata of the limestone which do not appear in the first part of the long route.

Their formation may be explained in this way: that portion of the rock where they are found consists of carbonate of lime, with some impurities, and contained originally the sulphide of iron, or iron pyrites, disseminated in small grains or crystals, and also in rounded nodules or concretions, sometimes of considerable size.

By exposure to air and moisture, oxygen unites with both the sulphur and the iron, producing sulphuric acid and oxide of iron, which combined, form a sulphate of iron. Then a double decomposition takes place; the sulphuric acid unites with the lime to form the gypsum; the carbonic acid of the limestone combines with the oxide of iron, forming a carbonate of iron, and this, on further exposure, parts with the carbonic acid, and leaves the brown coating of oxide, which is seen in many places on the surface of the rock.

The gypsum is thus constantly forming in the rock, and, being soluble, is carried by the water to the exposed surface where it crystallizes.

The crystals appear to grow out from the rock by additions from beneath, which continue to push the ends first formed, and if these do not become attached to other parts of the rock, straight needlelike fibres are often produced. Very commonly, however, the crystals begin to form when a small nodule of the iron ore is exposed at the surface; the parts first formed become attached to the surface around the edges, and as the chemical action proceeds towards the centre of the nodules successive leaf-like layers are thrown out, and the rosette form is the result. Along lines of fracture in the surface of the rock, the crystals are curved in opposite directions.

The wreaths and other figures formed by the chains of the rosettes, may be caused by the chemical action described taking place around the edges of large masses or concretions of the iron ore.

These crystalline forms occur only in the dryer parts of the cave. Where there is more moisture, as in the ' Snow-ball room,' the gypsum merely forms white, rounded concretions, originating from nodules of the iron ore on the roof and sides of the cave.'"

With these general remarks on the cave we give a brief account of its interesting fauna,* comprising representatives of the Fishes, Insects, and Crustaceans. No Mollusks nor Radiates have as yet been discovered, but the lower forms of life have been detected by Tellkampf, who collected several species belonging to the genera *Monas*, *Chilomonas*, and (?) *Chilodon*.

---

* In the following pages it will be noticed that the authors have expressed widely different views as to the origin of the peculiar forms of subterranean animals.

CHAPTER II.

THE CRUSTACEANS AND INSECTS.*

BY A. S. PACKARD, JR.

REPRESENTATIVES.of all the grand divisions of the Insects and Crustaceans have been found in this cave, and if no worms have yet been detected, one or more species would undoubtedly reward a thorough search.

We will enumerate what have been found, beginning with the higher forms. No Hymenoptera (bees, wasps, and ants) or Lepidoptera (moths) are yet recorded as being peculiar to caves. The Diptera (flies) are represented by two species, one of Anthomyia (Fig. 122) or a closely allied genus, and the second belonging to the singular and interesting genus Phora (Fig. 123). The species of Anthomyia usually frequent flowers; the larvæ live in decaying vegetable matter, or, like the onion fly, attack healthy roots; while the maggots of Phora live in decaying substances. It would be presumptuous in the writer to attempt to describe these forms without collections of species from the neighborhood of the cave, for though like all the rest of the insects they were found three or four miles from the mouth, yet they may be found to occur outside of its limits, as the eyes and the colors of the body are as bright as in other species.

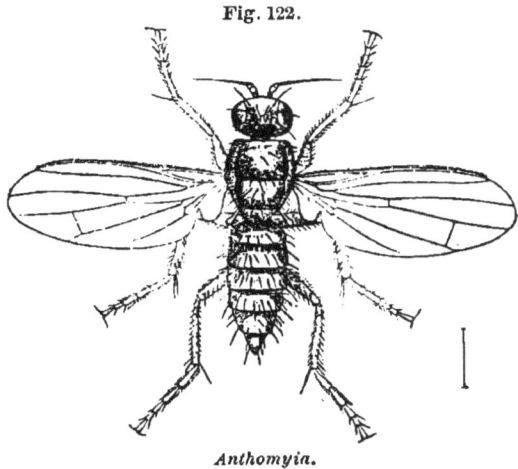

Fig. 122.

Anthomyia.

Among the beetles, two species were found by Mr. Cooke. The

*From the AMERICAN NATURALIST for December, 1871.

Anopthalmus Tellkampfii of Erichson, a Carabid (Fig. 124), and *Adelops hirtus* Tellkampf (Fig. 125) allied to Catops, one of the Silphidæ or burying beetle family. The Anopthalmus is of a pale reddish horn color, and is totally blind;[*] in the Adelops, which is grayish brown, there are two pale spots, which may be rudimentary eyes, as Tellkampf and Erichson suggest. No Hemiptera (bugs) have yet been found either in the caves of this country or Europe. Two wingless grasshoppers (generally called crickets) like the common species found under stones (*Ceuthophilus maculatus* Harris), have been found in our caves; one is the *Hadenœcus subterraneus* (Fig. 126 nat. size) described by Mr. Scudder, and very abundant in Mammoth Cave. The other species is *C. stygia* Scudder, from Hickman's cave, near Hickman's landing, upon the Kentucky river. It is closely allied to the Mammoth Cave species. According to Mr. Scudder, the specimens of *C. stygia* were found by Mr. A. Hyatt "in the remotest corner of Hickman's Cave, in a sort of a hollow in the rock, not particularly moist, but having only a sort of cave dampness. They were found a few hundred feet from the sunlight, living exclusively upon the walls." Even the remotest part of that cave is not so gloomy but that some sunlight penetrates it.

Fig. 124.

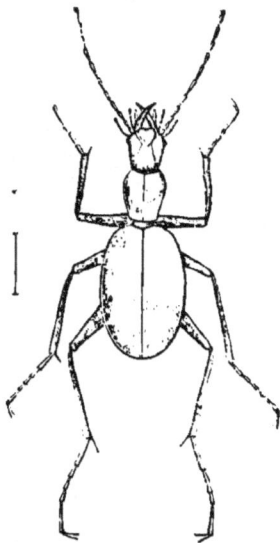

Anophthalmus Tellkampfii.

Fig. 123.

Phora.

Fig. 125.

Adelops hirtus.

The other species is found both in Mammoth Cave, and in the adjoining White's Cave. It is found throughout the cave, and most commonly (to quote Mr. Scudder) "about 'Martha's Vineyard' and in the neighborhood of 'Richardson's Spring' where they were discovered jumping about with the greatest alacrity upon the walls, where only they are found, and even when dis-

---

[*] In Erhardt's cave, Montgomery Co., Virginia, Prof. Cope found "four or five specimens of a new Anopthalmus, the *A. pusio* of Horn, at a distance of not more than three hundred feet from its mouth. The species is small, and all were found together under a stone. Their movements were slow, in considerable contrast to the activity of ordinary Carabidæ." Proc. Amer. Phil. Soc. 1869. p. 178.

turbed, clinging to the ceiling, upon which they walked easily; they would leap away from approaching footsteps, but stop at a cessation of the noise, turning about and swaying their long antennæ in a most ludicrous manner, in the direction whence the disturbance had proceeded; the least noise would increase their tremulousness, while they were unconcerned at distant motions, unaccompanied by sound, even though producing a sensible current of air; neither did the light of the lamp appear to disturb them; their eyes, and those of the succeeding species (*R. stygia*) are perfectly formed throughout, and they could apparently see with ease, for they jump away from the slowly approaching hand, so as to necessitate rapidity of motion in seizing them."

Mr. Henry Edwards has discovered a wingless grasshopper in a limestone cave at Collingwood, Massacre Bay, Middle Island, New

Fig. 126.

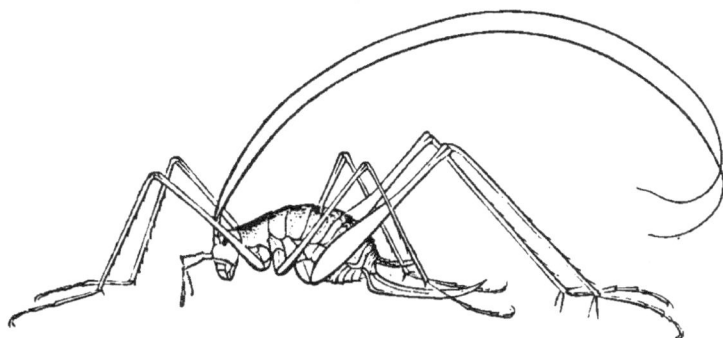

Hadenœcus subterraneus.

Zealand. Says Mr. Scudder, who has described the species in the "Proceedings of the Boston Society of Natural History" (Vol. xii, 1869, p. 408) under the name *Hadenœcus Edwardsii*, "the cave is close to the sea shore, and near a very large coal deposit, which occasionally crops out in the interior. The Hadenœci were rather numerous, but very difficult to catch, disappearing in the crevices of the rocks on the approach of lights. They appeared to be most abundant near the streams of water which percolated through the rocks." The wingless grasshopper of the European caves is the *Hadenœcus palpatus* Scudder, first described by Sulzer under the name *Locusta palpata*.

The Thysanurous Neuroptera are represented by a species of Machilis, allied to our common *Machilis variabilis* Say, common in Kentucky and the middle and southern states. So far as Tellkampf's figure indicates, it is the same species apparently, as I have received numerous specimens of this widely distributed form from Knoxville, Tennessee, collected by Dr. Josiah Curtis.

It was regarded as a crustacean by Tellkampf, and described under the name of *Triura cavernicola.** He mistook the labial and maxillary palpi for feet and regarded the nine pairs of abdominal spines as feet. The allied species, *M. variabilis* Say, is figured in vol. v. pl. 1, fig. 8, 9 (see also p. 94 of vol. v of the NATURALIST).

An interesting species of Campodea† of which the accompanying cut (Fig. 127) is a tolerable likeness, though designed to illustrate another species (*C. staphylinus* Westw.) was discovered by Mr. Cooke. Both the European and our common species live under stones in damp places, and the occurrence of this form in the water is quite remarkable. The other species are blind, and I could detect no eyes in the Mammoth Cave specimen.

A small spider was captured by Mr. Cooke, but afterwards lost; it was brown in color, and possibly distinct from the *Anthrobia monmouthia* Tellkf. (Fig. 128) which is an eyeless form, white and very small, being but half a line in length. The family

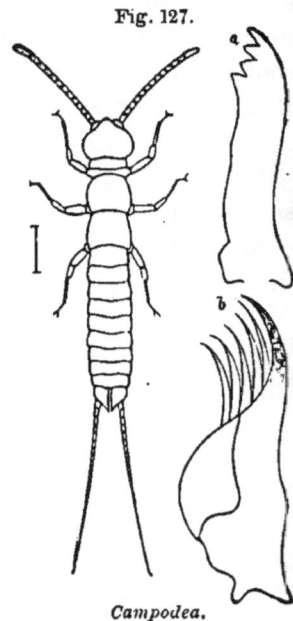

Fig. 127.

Campodea.

---

*Professor Agassiz in his brief notice of the Mammoth Cave animals, does not criticise Tellkampf's reference of this animal to the crustacea; and so eminent an authority upon the articulates as Schiödte remarks that while "Dr. Tellkampf's account affords us no means of forming any conclusion as to its proximate relations," that, however, it "appears to belong to the order of Amphipoda, and to have a most remarkable structure." Tellkampf's figure of Machilis is entirely wrong in representing the labial and maxillary palpi as ending in claws, thus giving the creature a crustacean aspect; and indeed he describes them as true feet!

† *Campodea Cookei* n. sp. Closely allied to *C. Americana*, but it is much larger; the antennæ are 24-jointed instead of 20-jointed as in *C. Americana*, and reach to the basal abdominal segment, while in *C. Americana* they reach only to the second thoracic; the terminal joints are much longer than in that species, the penultimate joint being one-third longer. Last three abdominal segments unequal (equal in *C. Americana*) the penultimate very short, not half as long as the terminal, which is longer and slenderer than in *C. Americana*, while the three are much narrower in proportion to the rest of the body than in the other species. Hind femora longer than in *C. Americana*. Entirely white and pilose. Length .25 inch, the largest *C. Americana* being .15 to .20 inch. (Anal stylets broken off.) Several specimens were seen by Mr. C. Cooke, but only one was captured in a pool of water, two or three inches deep, in company with the Cæcidotea.

of Harvest men is represented by a small white form, described by Tellkampf under the name of *Phalangodes armata* (Fig. 129) but now called *Acanthocheir armata* Lucas. The body alone is but half a line long, the legs measuring two lines. It should be borne in mind that many of the spiders, as well as the Thysanura, live in holes and dark places, so that we would naturally find them in caves. So, also, with the Myriopods, of which a most remarkable

Eig. 128.

Anthrobia monmouthia.

form * (Figs. 130, and 130 a front of head) was found by Mr. Cooke, three or four miles from the mouth of the cave. It is the only truly hairy species known, an approach to it being found in *Pseudotremia Vudii* Cope. It is blind, the other species of this group which Professor Cope found living in caves having eyes. The long hairs arranged along the back, seem to suggest that they are tactile organs, and of more use to the Thousand legs in making its way about the nooks and crannies of a perpetually dark cave than eyes would be. It was found by Mr. Cooke under a stone.

Prof. Cope has contributed to the " Proceedings of the American Philosophical Society" (1869, p. 171) an interesting account of the

---

* *Spirostrephon* (*Pseudotremia*) *Copei* n. sp. Head with rather short, dense hairs; no eyes, and no ocular depression behind the antennæ, the surface of the epicranium being well rounded to the antennal sockets; behind the insertion of the antennæ the sides of the head are much more swollen than in *S. lactarius.* Antennæ slender, with short thick hairs; relative length of joints, the 6th being longest; 6th, 4th, 5th, 3d, 8th, 7th, 1st, the 7th joint being much thicker than the 8th. Twenty-eight segments besides the head; they are entirely smooth, striated neither longitudinally nor transversely; a few of the anterior segments rapidly decrease in diameter towards the head. The segments are but slightly convex, and on each side is a shoulder, bearing three tubercles in a transverse row. each giving rise to a long stiff hair one-half to two-thirds as long as the segment is thick; these hairs stand up thickly all over the back, and may serve at once to distinguish the species. No pores. Feet long and slender, nearly as long as the antennæ, being very slender towards the claws. Entirely white. Length of body .35 inch; thickness .04 inch.

It is nearly allied to *Pseudotremia Vudii* of Cope. It will be noticed that Professor Cope characterizes the genus Spirostrephon as having "no pores"; though we find it difficult to reconcile this statement with that of Wood who describes *S. lactarius* as having "lateral pores." Cope separates Pseudotremia from Spirostrephon for the reason that the segments have "two pores on each side the median line." The present species has no pores, but seems in other characters to be a true Spirostrephon, and we are thus led to doubt whether Pseudotremia is a well founded genus.

cave mammals, articulates and shells of the middle states.   He
says that " myriopods are the only articulates which can be
readily found in the remote regions of the caves [of West Vir-
ginia] and they are not very common in a living state."   The
Pseudotremia cavernarum which he describes, " inhabits the deep-

Fig. 129.

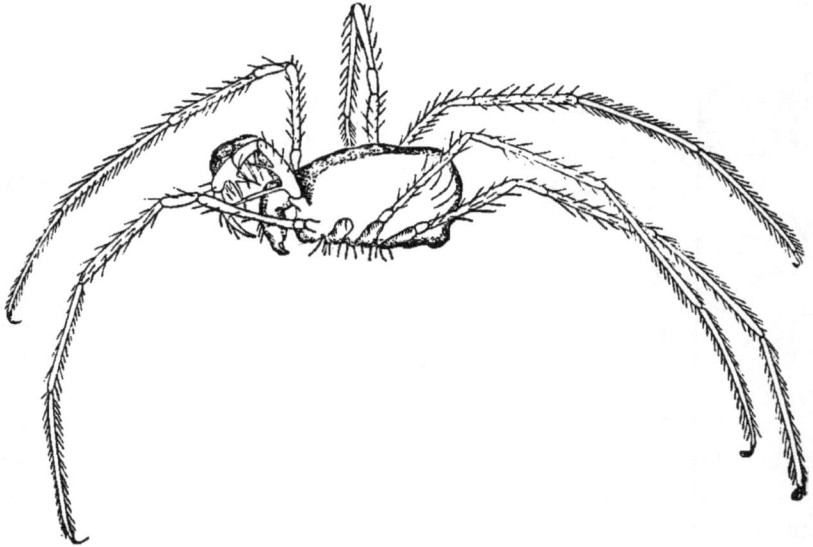

Acanthocheir armata.

est recesses of the numerous caves which abound in Southern Virginia, as far as human steps can penetrate.   I have not seen it
near their mouths, though its eyes are not undeveloped, or smaller
than those of many living in the forest.   Judging from its remains,
which one finds under stones, it is an abundant species, though

Fig. 130.   Fig. 130 a.

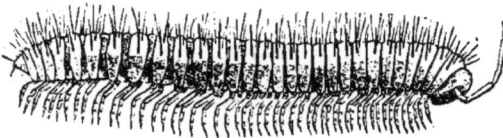

Spirostrephon Copei.

rarely seen by the dim light of a candle even after considerable
search.   Five specimens only were procured from about a dozen
caves."   The second species, *P. Vudii* Cope, was found in Montgomery Co. and he thinks it was not found in a cave.   Professor
Hyatt informs me that he saw near the "Bottomless Pit" in Mam-

moth Cave, a brownish centipede-like myriopod, over an inch in length, which moved off in a rapid zigzag motion. Unfortunately he did not capture it.

Next to the blind fish, the blind crawfish attracts the attention of visitors to the cave. This is the *Cambarus pellucidus* (Fig.

Fig. 121.

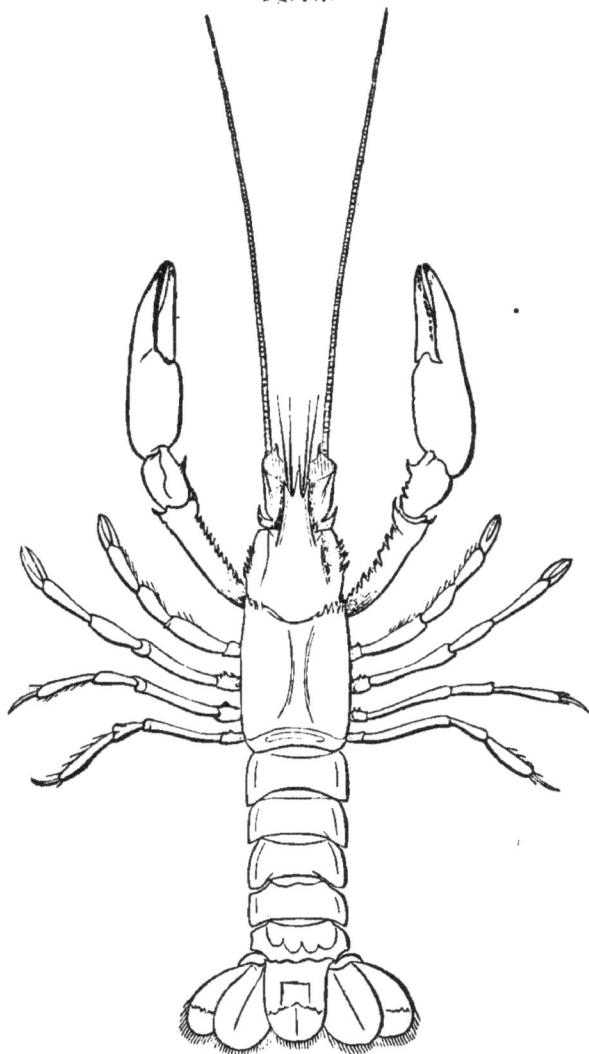

Cambarus pellucidus.

131, from Hagen's monograph of the North American Astacidæ) first described by Dr. Tellkampf. He remarks that "the eyes are rudimentary in the adults, but are larger in the young." We might add that this is an evidence that the embryo develops like those of the other species; and that the inheritance of the blind condition is probably due to causes first acting on the a-dults and transmitted to their young, until the production of offspring that become blind becomes a habit. This is a partial proof at least that the characters separating the genera and species of animals are

those inherited from adults, modified by their physical surroundings and adaptations to changing conditions of life, inducing certain alterations in parts which have been transmitted with more or less rapidity, and become finally fixed and habitual. Prof. Hagen has seen a female of *Cambarus Bartonii* from Mammoth Cave, " with the eyes well developed," and a specimen was also found by Mr. Cooke. Prof. Hagen remarks that " *C. pellucidus* is the most aberrant species of the genus. The eyes are atrophied, smaller at the base, conical, instead of cylindrical and elongated, as in the other species. The cornea exists, but is small, circular, and not faceted ; the optic fibres and the dark-colored pigments surrounding them in all other spe-

cies are not developed." It seems difficult for one to imagine that our blind craw fish was created suddenly, without the intervention of secondary laws, for there are the eyes *more perfect in the young than the adult*, thus pointing back to ancestors unlike the species now existing. We can now understand, why embryologists are anxiously studying the embryology of animals to see what organs or characteristics are inherited, and what originate *de novo*, thus building up genealogies, and forming almost a new department of science : comparative embryology in its truest and widest sense.

Of all the animals found in caves, either in this country or Europe, perhaps the most strange and unexpected is the little creature of which we now speak. It is an Isopod crus-

Fig. 132.

Cæcidotea stygia (side view).

Fig. 133.

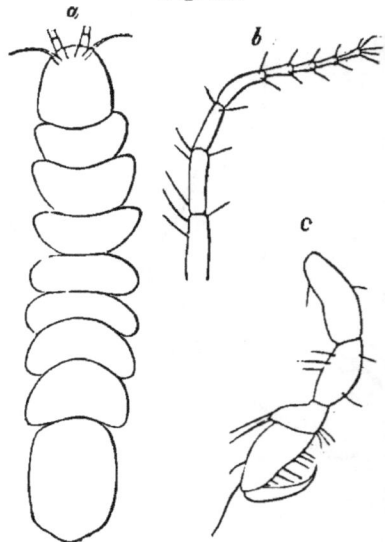

Cæcidotea stygia (dorsal view).

tacean, of which the pill bugs or sow bugs are examples. A true species of pill bug (*Titanethes albus* Schiödte) inhabits the caves of Carniolia, and it is easy to believe that one of the numerous species of this group may have become isolated in these caves and modified into its present form. So also with the blind *Niphargus stygius* of Europe, allied to the fresh water Gammarus so abundant in pools of fresh water. We can also imagine how a species of Asellus, a fresh water Isopod, could represent the Idoteidæ in our caves, and one may yet be found; but how the present form became a cave dweller is difficult of explanation, as its nearest allies are certain species of Idotea which are all marine, with the exception of two species: *I. entomon*, living in the sea and also in the depths of the Swedish lakes, as discovered by Loven, the distinguished Swedish naturalist, while a species representing this has been detected by Dr. Stimpson at the bottom of Lake Michigan. Our cave dweller is nearly allied to Idotea, but differs in being blind, and in other particulars, and may be called *Cæcidotea stygia.* * (Fig. 132 side view, enlarged; Fig. 133 dorsal view; *b*, inner antenna; *c*, 1st leg.) It was found creeping over the fine sandy bottom, in company with the Campodea, in a shallow pool of water four or five miles from the mouth of the cave.

This closes our list of known articulates from this and other caves in this country, the result of slight explorations by a few individuals. The number will be doubtless increased by future research. It is to be hoped that our western naturalists will thoroughly explore all the sinks and holes in the cave country of the western and middle states. The subject is one of the highest interest in a zoological point of view, and from the light it throws on the doctrine of evolution. Professor Schiödte, the eminent Danish zoologist, has given us the most extended account of the cave fauna of Europe, which has been translated from the Danish into the Transactions of the Entomological Society of London (new series vol. 1, 1851).

He examined four caves; namely, that of Adelsberg, the Magdalena and Luege caves, all in the neighborhood of Adelsberg,

---

* Generic characters. Head large, much thicker than the body, and as long as broad; subcylindrical, rounded in front. No eyes. First antennæ slender, 8-jointed (2d antennæ broken off). Abdominal segments consolidated into one piece. Differs chiefly from Idotea, to which it is otherwise closely allied, by the 8-jointed (instead of 4-jointed) 1st (inner) antennæ, the very large head, and by the absence of any traces of the three basal segments of the abdomen usually present in Idotea.

Specific characters. Body smooth, pure white: tegument thin, the viscera appearing through. Head as wide as succeeding segment, and a little more than twice as long. Inner antennæ minute, slender, the four basal joints of nearly equal length, though the fourth is a little smaller than the basal three, remaining four joints much smaller than others, being one-half as thick and two-thirds as long as either of the four basal joints; ends of last four joints a little swollen, giving rise to two or three hairs; terminal joint ending in a more distinct knob, and bearing five hairs. Segment of thorax very distinct, sutures deeply incised; edges of segments pilose; abdomen flat above, rounded behind, with a very slight median projection; the entire pair of gills do not reach to the end of the abdomen, and the inner edges diverge posteriorly. Legs long and slender, 1st pair shorter, but no smaller than the second. Length .25 inch.

and the Corneale cave at Trieste.   The only plant found was a
sort of fungus, *Byssus fulvus* Linn.   The only vertebrate is the
singular salamander, *Hypochthon* (Proteus) *anguinus*, found in the
Magdalina river.   No shells were found.   Regarding the articulates he writes: ·

"On searching along the walls within the entrance of the caves,
among the rubbish and the vegetable debris along the sides of the
river, we meet with a considerable number of Insecta, Myriopoda,
Arachnida and Crustacea, of various families which shun daylight;
being such species only as inhabit promiscuously other places,
provided they are moist and feebly illumined.   We find species of
Pterostichus, Pristonychus, Amara, Quedius, Homalota, Omalium,
Hister, Trichopteryx, Cryptophagus, Atomaria, Ptinus, Ceraphron,
Belyta, a grasshopper of the Locust family, probably the *Raphidophora cavicola* Fischer, as it was only seen in the larva state, Trich-
optera, Sciara, Psychoda, Phora, Heteromyza, Sapromyza, Tomoce-
rus, Linyphia, Gamasus, Cryptops, Julus, and Asellus.   In proportion as we recede from the entrance the number of species as
well as individuals greatly decreases, and at the distance which
entirely excludes the light, only single individuals are found.   In
the deepest recesses these species are entirely wanting, except
some few which have been transported by the current; only a few
Diptera are found; namely, a species of Phora, very near *P. ma-
culata* Meig., *Heteromyza flavipes* Zett., and *Sapromyza chrysoph-
thalma* Zett., extending also very far into the caves, even to the
remotest accessible places in Adelsberg cave, more than half an
hour's walk from its entrance.   Dead moths are occasionally found
far in the caves, being left there by the bats; and likewise accidental specimens of the parasites of the latter.   Of the five earlier known animals which inhabit these caves, I found *Pristonycha
elegans* Dej. rather frequently, and *Homalota spelæa* Er. in considerable numbers.   Besides these are *Anopthalmus Schmidtii*, which
is very rare, and the wood louse, *Titanethes alba*.   The new forms
he found were a beetle (*Bathyscia byssina*) allied to our Adelops;*
Stagobius troglodytes, an aberrant genus of Silphids; a Podurid,
Anurophorus Stillicidii; and the two blind arachnidans, one a spider allied to Dysdera, the *Stalita tænaria*, and a false-spider, *Blothrus spelæus*.   Among the crustacea he found *Niphargus stygius*, †

---

* Ludwig Müller enumerates four other species of Adelops from these caves, and
three species from France, and *Machærites spelœus*, in Verhandl. Zool. Bot. Vereins.
Wien. 1855, p. 505. See also Heller's Beitrage zur Österreich. Grotten-Fauna. (Myriopoda and Crustacea.) Vienna. 1858. He describes a myriopod with rudimentary eyes
(*Trachysphæria Schmidtii*) allied to Glomeris, and another blind species (*Brachydesmus
subterraneus*) allied to Polydesmus; also a new *Tithanethes* (*T. graniger*), and notices
Monolistra cœca Gerst. Waukel (1861) also found a new Phalangid (*Leiobunum troglodytes*) with distinct eyes and four species of mites in the caves of Eastern Austria. The
mites are *Scyphius spelœus, Linopodes subterraneus, Gamasus loricatus* and *G. nireus*,
and an additional species of Trachysphæria (*T. Hyrtlii*). See also Ehrenberg's list of
cave insects (Monabsberichte der Akad. Berlin. 1861.)
† Several species of Niphargus occur in the wells and hot springs in Europe. Accord-

allied to Gammarus, which lives in small pools of water and is white and blind ; and the cave pill bug, *Titanethes albus* (Koch.).''

In conclusion Schiödte remarks that : —

"We may with propriety apply the collective term *Subterranean Fauna* to those animals which exclusively inhabit caves, and are expressly constructed for such habitations. Still there is nothing in this name which would indicate that these animals have any claim to be considered as a separate group, beyond the mere peculiarity of their common place of abode. While a few of them possess such an extraordinary structure as to stand in no comparison with those animals which inhabit the light, there are others, forming only more characteristic links in the groups of animals more or less shy of light, of which many are found common in the localities of the caves ; and some belong to genera having a wide local, as well as geographical, extension. We are accordingly prevented from considering the entire phenomenon in any other light than something purely local, and the similarity which is exhibited in a few forms (Anophthalmus, Adelops, Bathyscia) between the Mammoth Cave and the caves in Carniola, otherwise than as a very plain expression of that analogy, which subsists generally between the fauna of Europe and of North America. Besides, it is clear to me that the fauna of the caves of Carniola consists of two divisions, of which the essential character is referable on the one hand to the dark locality, and on the other to the additional confinement to stalactitic formations ; as yet we are not

---

ing to Bate and Westwood (British Sessile eyed Crustacea) " the British examples have been obtained from artificially excavated wells connected with houses for domestic purposes. In some instances the wells have been old, in others but recently dug. In their geological condition the habitats have been equally various. At Corsham the well exists in the Oolite formation, at Ringwood in chalk-flint gravel, at Mannamead in slate. At Corsham and Mannamead they are found on a hill, at Ringwood they lie low. The appearance of some of these animals in a well soon after its being excavated, raises a question of considerable interest. Thus they were found at upper Clafford, near Andover and at Mannamead, near Plymouth, but not a trace of them was to be found in the surrounding streams; in fact they perish in the light. It is impossible to regard them as an extreme variety, or modification of our only fresh water Amphipod, *Gammarus fluviatilus*, since various parts not only differ in form, but some are altered in character; for example, the extraordinary elongation and slenderness of one of the branches of each of the last pair of caudal appendages seem to be a special structure, having for its object the antenna-like use of a delicate apparatus at the extremity of the body. . . . . Although we can find no fresh water ally to this genus in the rivers and streams of Europe, yet Bruzelius has taken in the deep sea, near Bohusia, a form which he has described under the name *Eriopis elongata*, approximating so nearly to it that it appears to be scarcely generically distinct.

We are inclined to think that *Gammarus pungens* of Milne Edwards, from the warm springs of Cassini in Italy, also belongs to this genus." Of Niphargus these are the following species known besides *N. stygius*; i.e. *N. aquilex* Schiödte (*Gammarus puteanus* Koch, the embryology of which has been studied by V. St. George) *N. fontanus* Bate, *N. Kochianus* Bate. Another generic form is *Crangonyx* founded by Bate, which also belongs to the subterranean fauna. " A single species as yet is all that has been found in England; but we have little doubt but that *Gammarus Ermanni* of Milne Edwards which was found by M. Ermann in the warm springs of Kamtschatka belongs also to this genus. It is curious that we should have to record that while the animals of this genus, as in the preceding (Niphargus) inhabit the deep artificial wells, without being known to exist in our rivers and streams, its nearest allied form is to be found in a marine genus, Gammarella.''

able vigorously to discriminate between the two. We shall accordingly look upon the subterranean fauna, or more properly faunas, as small ramifications which have penetrated into the earth from the geographically-limited faunas of the adjacent regions; and which, as they extended themselves into darkness, have been accommodated to surrounding circumstances. Animals not far remote from the ordinary forms, prepare the transition from light to darkness. Next follow those that are constructed for twilight; and last of all those destined for total darkness, and whose structure is quite peculiar. Among these some are adapted for special localities, those which inhabit dry localities or detached little reservoirs being totally blind, while others, destined for running streams, have eyes of imperfect construction, so as to receive the impression of rays of light, but no proper image of illuminated objects. We may therefore with tolerable precision arrange the inhabitants of caverns under the following heads:—

Shade animals.—Extensive genera and species inhabiting caverns near their entrance, and, generally, all cool, shady and moist localities. Of these, those that fly occasionally enter far into the caverns (Diptera).

Twilight animals.—They belong to widely spread genera, but are peculiar to the caves, and distinguished by their small eyes. They are principally found near the entrances to the caves, but proceed deeper into the darkness than the shade-animals, and although wingless, they penetrate often the whole extent of the dark space.—(*Pristonychus elegans, Homalota spelæa.*)

Cave animals.—They form, at least in part, peculiar genera, are wingless and colorless, as far as the consistency of their integuments will admit, and exist exclusively in total darkness. The terrestrial division is blind; the aquatic has a perception of light. To this group belong all the animals in the Mammoth Cave, and among those of the caves of Carniola, Anophthalmus, Bathyscia, perhaps likewise Anurophorus and Hypochthon, which, however, may belong to the following group.

Stalactite cave animals.—Insects, Arachnidans and Crustaceans appertaining to peculiar genera, wingless, blind, brightly colored according to the nature of their integuments, either light brown, yellowish white, or snow white, perhaps according to the preponderance of the chitine; living in total darkness, peculiar to stalactite caves, in part occupying the columns and constructed accordingly, either for ascent or hovering over them. Here belong most of the animals treated of in this memoir—Stagobius, Blothrus, Stalita, Niphargus, and Titanethes."[*]

A pertinent question arises as to the time of the formation of these caves and when they became inhabitable. As previously stated, the caves of the western and middle States are in lower Carboniferous limestone rocks, though the Port Kennedy cave explored by Wheatley and Cope† is in the Potsdam limestone. They could not have been formed under water, but when the land was drained' by large rivers. This could not have occurred previous to the Triassic period. Prof. Dana in his "Manual of Geology" shows that the Triassic continent spread westward from the Atlantic coast "to Kansas, and southward to Alabama; for through this great area there are no rocks more recent than the Palæozoic." "Through the Mesozoic period [comprising the Triassic, Jurassic, and Cretaceous periods] North America was in general dry land, and on the east it stood a large part of the time above its present level." Though at the close of these periods there was a general extinction of life, yet this was not probably a sudden (one of months and even years), but rather a secular extinction, and there may be plants and animals now living on dry land, which are the lineal descendants of mesozoic and more remotely of Carboniferous forms of life. So our cave animals may possibly be the survivors of Mesozoic forms of life, just as we find now living at great depths in the sea remnants of Cretaceous life. But from the recent explorations in the caves of Europe and this country, especially the Port Kennedy cave, with its remarkable assemblage of vertebrates and insects, we are led to believe from the array of facts presented by Prof. Cope that our true subterranean fauna probably does not date farther back than the beginning of the Quaternary, or Post pliocene, period. We quote his "general observations" in his article on the Port Kennedy fauna.

"The origin of the caves which so abound in the limestones of the Alleghany and Mississippi valley regions, is a subject of much interest. Their galleries measure many thousands of miles, and their number is legion. The writer has examined twenty-five, in more or less detail, in Virginia and Tennessee, and can add his testimony to the belief that they have been formed by currents of running water. They generally extend in a direction parallel to the strike of the strata, and have their greatest diameter in the direction of the dip. Their depth is determined in some measure by the softness of the stratum, whose removal has given them existence, but in thinly stratified or soft material, the roofs or large

† A notice of the animals found in this cave will be found in the Proceedings of the American Philosophical Society, April, 1871. The insects there enumerated would probably not come under the head of cave insects.

masses of rocks fall in, which interrupt the passage below. Caves, however, exist when the strata are horizontal. Their course is changed by joints or faults, into which the excavating waters have found their way.

That these caves were formed prior to the postpliocene fauna is evident from the fact that they contain its remains. That they were not in existence prior to the drift is probable, from the fact that they contain no remains of life of any earlier period so far as known, though in only two cases, in Virginia and Pennsylvania, have they been examined to the bottom. No agency is at hand to account for their excavation, comparable in potency and efficiency to the floods supposed to have marked the close of the glacial period, and which Prof. Dana ascribes to the Champlain epoch. An extraordinary number of rapidly flowing waters must have operated over a great part of the Southern States, some of them at an elevation of fifteen hundred feet and over (perhaps two thousand) above the present level of the sea. A cave in the Gap Mountain, on the Kanawha river, which I explored for three miles, has at least that elevation.

That a territory experiencing such conditions was suitable for the occupation of such a fauna as the deposits contained in these caves reveal, is not probable. The material in which the bones occur in the south is an impure limestone, being mixed with and colored by the red soil which covers the surface of the ground. It is rather soft but hardens on exposure to the air.

The question then remains so far unanswered as to whether a submergence occurred subsequent to the development of the postpliocene mammalian fauna. That some important change took place is rendered probable by the fact, that nearly all the neotropical types of the animals have been banished from our territory, and the greater part of the species of all types have become extinct. Two facts have come under my observation which indicate a subsequent submergence. A series of caves or portions of a single cave once existing on the southeast side of a range of low hills among the Alleghany mountains in Wythe Co., Virginia, was found to have been removed by denudation, fragments of the bottom deposit only remaining in fissures and concavities, separated by various intervals from each other. These fragments yielded the remains of twenty species of postpliocene mammalia.* This denudation can be ascribed to local causes, following a subsidence of uncertain extent. In a cave examined in Tennessee the ossiferous deposit was in part attached to the *roof* of the chamber. Identical fossils were taken from the floor. This might, however, be accounted for on local grounds. The islands of the eastern part of the West Indies appear to have been separated by submergence of larger areas, at the close of the period during which they

---

* See Proceed. Amer. Phil. Soc. 1869, 171.

were inhabited by postpliocene mammalia and shells. The caves of Anguilla include remains of twelve vertebrates,* of which seven are mammalia of extinct species, and several of them are of large size. These are associated with two recent species of molluscs *Turbo pica*, and a *Tudora* near *pupæformis*.† As these large animals no doubt required a more extended territory for their support than that represented by the small island Anguilla, there is every probability that the separation of these islands took place at a late period of time and probably subsequent to the spread of the postpliocene fauna over North America."

I think the reader will conclude from the facts Prof. Cope so clearly presents, that the subterranean fauna of this country does not date back of the Quaternary period. These species must have been created and taken up their abode in these caves (Mammoth Cave and those of Montgomery County, Virginia) after the breccia flooring their bottoms and containing the bones of Quaternary animals had been deposited; or else migrated from Tertiary caves farther south, which is not probable, as it has been previously shown that those blind animals inhabiting wells immediately die on being exposed to the light (British Sessile-eyed Crustacea, i, p. 313), though the blind craw fish is not thus affected.

The case becomes much simpler when we consider the age of the rocks in which the Adelsberg and other caves mentioned by Schiödte are situated. The Alps were under water in the Middle Eocene; consequently the caves could not have been formed until the close of the Tertiary. Hence the species of the cave fauna were evidently created either at the close of the Tertiary, or more probably the beginning of the Quaternary, as "even in the later part of the Pliocene era there was an elevation of three thousand feet in a part of the Island of Sicily" (Dana). We are therefore led to conclude that the species of the subterranean fauna the world over are recent creations, probably not older than the extinct mammals associated with man.

---

* Loc. cit. 1869, 183; 1870, 608. A fourth species of gigantic Chinchillid has been found by Dr. Rijgersma, which may be called *Loxomylus quadrans* Cope. It is represented by portions of jaws and teeth of three individuals. It is one of the largest species, equalling the *L. latidens*, and has several marked characters. Thus the roots of the molars are very short, and the triturating surface oblique to the shaft. The roots of the second and fourth are longer than those of the first and third. The last molar has four dental columns instead of three as in the other *Loxomyli*, and is triangular or quadrant-shaped in section; the third is quadrangular in section, and has three columns. The second is the smallest, being only .6 the length of the subtriangular, first. Length of dental series m .063 or 2.5 inches. Palate narrow and deeply concave. There is but little or no lateral constriction in the outlines of the teeth; the shanks are entirely straight. In its additional dentinal column, this species approaches the genus *Amblyrhiza*.

The large Chinchillas of Anguilla are as follows, *Loxomylus longidens*, *L. latidens*, *L. quadrans*, and *Amblyrhiza inundata*.

† See Bland, Proceed. Amer. Phil. Soc., 1871, 58.

Assuming on the principles of evolution that the cave animals were derived from other species changed by migration from the outer world to the new and strange regions of total darkness, it seems evident that geologically speaking the species were *suddenly* formed, though the changes may not have been wrought until after several thousand generations. According to the doctrine of natural selection, by which species pass from one into another by a great number of minute variations, this time was not sufficient for the production of even a species, to say nothing of a genus. But the comparatively sudden creation of these cave animals affords, it seems to us, a very strong argument for the theory of Cope and Hyatt of creation by acceleration and retardation, which has been fully set forth in this journal. The strongly marked characters which separate these animals from their allies in the sunlight, are just those fitting them for their cave life and those which we would imagine would be the first to be acquired by them on being removed from their normal habitat.

On introducing the wingless locust, *Ceuthophilus maculatus*, into a cave, where it must live not under stones, but by clinging to the walls, its legs would tend to grow longer, its antennæ and palpi would elongate and become more delicate organs of hearing as well as touch,* and the body would bleach partially out, as we find to be the case in *H. subterranea* and *C. stygia.* The Carabid beetle, Anopthalmus, extending farther into the cave, would lose its wings (all cave insects except the Diptera have no wings, elytra excepted) and eyes, but as nearly all the family are retiring in their habits, the species hiding under stones, its form would not undergo farther striking modification. So with the blind Campodea, which does not differ from its blind congeners, which live more or less in the twilight, except in its antennæ becoming longer. The blind Adelops, but with rudiments of eyes, does not greatly depart in habits from Catops, while on the other hand the remarkable Stagobius of the Illyrian caves, which according to

---

* After writing this article. and without knowledge of his views, we turned to Darwin's Origin of Species to learn what he had to say on the origin of cave animals. He attributes their loss of sight to disuse, and remarks :—" By the time an animal has reached, after numberless generations, the deepest recesses, disuse will on this view have more or less perfectly obliterated its eyes, and natural selection will often have effected other changes, such as an increase in the length of the antennæ or palpi, as a compensation for blindness." 5th Amer. Edit., p. 143. We are glad to find our views as to the increase in the length of the antennæ and palpi compensating for the loss of eyesight, confirmed by Mr. Darwin.

Schiödte spends its life in crawling ten to twenty feet above the floor over the columns formed by the stalactites, to which unique mode of life it is throughout perfectly adapted, is remarkably different from other Silphids. Its legs are very long and inserted far apart (the prothorax being remarkably long), with surprisingly long claws, while the antennæ, again, are of great length and densely clothed with hairs, making them most delicate sense organs.* So also are the limbs of the false scorpion, and the spider and pill bug (Titanethes) of remarkable length.

But the modifications in the body of the Spirostrephon are such that many might deem its aberrant characters as of generic importance. It loses its eyes, which its nearest allies in other, but smaller, caves possess, and instead gains in the delicate hairs on its back, which evidently form tactile organs of great delicacy; the feet are remarkably long, as also the antennæ. These are not new formations but simply modifications, apparently by use or disuse, of organs present in the other species. The aberrant myriopod and Stagobius are paralleled by the blind fish, an animal so difficult to classify, and so evidently adapted for its abode in endless darkness. And as an additional proof of the view here taken that these cave animals are modified from more or less allied species existing outside of the caves, we have the case of the craw fish, whose eyes (like those of the mole), are larger in the young than adult, indicating its descent from a species endowed with the faculty of sight, while in the adult the appendages are modified as tactile organs so as to make up for its loss of eyesight, in order that it may still take its prey.

We thus see that these cave animals are modified in various ways, some being blind, others very hairy, others with long appendages. All are not modified in the same way in homologous organs; another argument in proof of their descent from ancestors

* Schiödte remarks that "it is difficult to understand the mode of life of *Stagobius troglodytes;* or how this slow and defenceless animal can escape being devoured by the rapid, piratical Arachnidans, or find adequate support on columns, for inhabiting which it is so manifestly constructed. We are led in this respect to consider the antennæ. Whatever significance we attach to those enigmatical organs, we must admit that they are organs of sense, in which view an animal having them so much developed as Stagobius, must possess a great advantage over its enemies, if these be only Arachnidans. Its cautious and slow progress, and its timid reconnoitring demeanor, fully indicate that it is conscious of life being in perpetual danger, and that it endeavors to the utmost to avoid that danger. Darkness, which always favors the pursued more than the pursuer, comes to its aid, especially on the uneven excavated surface of the columns."

whose habits varied, as those of their out-of-door allies do at present. Had they been specially created for subterranean life, we should have expected a much greater uniformity in the organs adapting them to a cave life than we actually find to be the case.

Another fact of interest in this connection is the circumstance that these cave species breed slowly, being remarkably poor in individuals ; they are nearly all extremely rare.* Did they breed as numerously as their allies in the outer world the whole race would probably starve, as the supply of food even for those which do live is wonderfully limited.

It is now known that animals inhabiting the abysses of the sea are often highly colored : light must penetrate there, for we know that were the darkness total they would be colorless like the cave insects.

In view of the many important questions which arise in relation to cave animals, and which have been too imperfectly discussed here, we trust naturalists the world over will be led to explore caves with new zeal, and record their discoveries with minuteness, and the greatest possible regard to exactness. The caves of the West Indian Islands should first of all be carefully explored. Also those of Brazil, those of the East Indies and of Africa, while fresh and most extended explorations of our own Mammoth Cave should be made, perhaps by a commission acting under government or State authority, in order that the most ample facilities may be afforded by the parties owning the cave.

NOTE.—Since my article was printed, Prof. Cope's article entitled " Life in the Wyandotte Cave" has appeared in the "Annals and Magazine of Natural History" (London) for November. He enumerates the following articulates as inhabitants of this cave; "Anophthalmus Tellkampfii, and another species; two species of Staphylinidæ; Raphidophora; two species of flies; an Aranea-like and Opilio-like spider; a species of Pseudotremia; Cambarus pellucidus, an unknown aquatic Crustacean with external egg pouches, and a Lernæan (crustacean) parasitic on the blind fish. Of these one beetle (Anophthalmus), the cricket (Raphidophora), a fly, the Opilio-like spider, the centipede, and the blind crawfish, are probably the same as those found in the Mammoth Cave. Two beetles and two crustaceans are certainly different from those of the latter, and the centipedes are much more numerous. The Gammaroid crustacean found in the waters of the Mammoth Cave, and which is, no doubt in part, the food of the blind fish, we did not find; but some such species no doubt exists, as we found an abundance of a lively little tetradecapod crustacean near the mouth of a cave close by."

* The wingless grasshoppers are common however, and Prof. Hagen writes me that the cave insects in Europe are probably not so rare as they are thought to be by naturalists, since the guides do not show the best collecting places, wishing to keep a stock on hand to sell to visitors.

CHAPTER III.

THE BLIND FISHES OF THE MAMMOTH CAVE AND THEIR ALLIES.*

BY F. W. PUTNAM.

THE blind fish of the Mammoth Cave has from its discovery been regarded with curiosity by all who have heard of its existence, while anatomists and physiologists have considered it as one of those singular animals whose special anatomy must be studied in order to understand correctly facts that have been demonstrated from other sources ; and, in these days of the Darwinian and development theories, the little blind fish is called forth to give its testimony, pro or con.

Before touching upon this point, however, we must call attention to the structure of the fish and its allies, and to others that are either partially or totally blind.

In the lancelet (Branchiostoma) and the hag (Myxine) the eye is described "as simple in form as that of a leach, consisting simply of a skin follicle † coated by a dark pigment, which receives the end of a nerve from the brain." Such an eye speck as this structure gives would only answer for the simple perception of light. In the young ‡ of the lampreys (Petromyzon) the eye is very small and

---

* From the AMERICAN NATURALIST for January, 1872.

† See further on where Prof. Wyman questions this structure.

‡ These young lampreys have been described under the generic name of Ammocœtes, and it was not until 1856, when Prof. Müller discovered the fact of a metamorphosis in the lampreys, that their true position was ascertained. Prof. Müller has traced the history of the common European species and shown that it is three or four years in attaining its perfect form. With this fact before us and with the early stages of the Myxinoids still unknown, have we not some reason for suspecting that the Lancelet may yet prove to be a larval form of the Myxinoids, notwithstanding that it is said to lay eggs ? Why should we not suspect the existence in the very lowest vertebrates of some-

placed in a fold of the skin of the head, and probably of little use, as these young remain buried in the sand; but as they attain maturity, and, with it, the parasitic habits of the adult, their eyes are developed to a fair size, thus reversing the general rule in the class.

In most other fishes the eyes are developed to a full and even remarkable extent as to size and perfection of sight in water. In Anableps, or the so called four eyed fish of the fresh waters of Central and South America, which belongs to a closely allied family with our blind fish, the *Cyprinodontidæ*, the eyes are not only fully developed, but are divided into an upper and lower portion in such a way, by an opaque horizontal line, as to give the effect of two pupils, by which the fish probably sees as well when following its prey on the surface with its eyes out of water, as when under water. But it is in the interesting family of cat fishes (*Siluridæ*) that we find the most singular arrangement of eyes in perfect adaptation to the diversified modes of life of the numerous species. In this family the eyes assume nearly every possible modification from partial and even total blindness to perfectly developed eyes, and these organs are placed in almost every conceivable position in a fish's head; from the ordinary large eyes on the side, to small ones on top of the head, enabling the fish to see only what is above; to the oval eyes on the side, in some just back of the mouth, situated in such a way that the fish can only see what is in close proximity to its jaws or even below them. Many genera of this family found in South America,* Africa† and Asia,‡ have the eyes so small and buried under the skin or protected by folds or cartilage, as evidently to be of no more use than simply to distinguish light from darkness.

Among the most interesting forms of this family, in this respect, is the genus described by Prof. Cope under the name of *Gronias nigrilabris*. This fish is very closely allied to our common bull

---

thing akin to "alternate generation." or of larvæ capable of reproduction? Without having any facts to support such an assumption, except that, on general principles, the young of Myxine would probably be very much like Branchiostoma, and that its young is not known, while Branchiostoma has only been found in waters where some species of Myxinoid exists, I think that before the position of the lancelet is firmly established we must know the embryology of the Myxinoids; for should the lancelet prove not to be the young of the Myxinoids, it must necessarily form a distinct class of animals, perhaps as near to the mollusks as to the vertebrates.

* *Pimelodus cyclopium* of Humboldt, *Helogenes*, *Agoniosus* and other genera.

† *Eutropius congensis.*

‡ *Ailia*, *Shilbichthys*, *Bagroides* and other genera.

pout or horned pout, and of about the same size (ten inches in length). It was taken in the Conestoga river in Lancaster Co., Penn., where it is "occasionally caught by fishermen and is supposed to issue from a subterranean stream said to traverse the limestone in that part of Lancaster Co., and discharge into the Conestoga." We quote the following from Prof. Cope's remarks on the fish : * —

"Two specimens of this fish present an interesting condition of the rudimental eyes. On the left side of both a small perforation exists in the corium, which is closed by the epidermis, representing a rudimental cornea ; on the other the corium is complete. Here the eyeball exists as a very small cartilaginous sphere with thick walls, concealed by the muscles and fibrous tissue attached, and filled by a minute nucleus of pigment. On the other the sphere is larger and thinner walled, the thinnest portion adherent to the corneal spot above mentioned ; there is a lining of pigment. It is scarcely collapsed in one, in the other so closely as to give a tripodal section. Here we have an interesting transitional condition in one and the same animal, with regard to a peculiarity which has at the same time physiological and systematic significance, and is one of the comparatively few cases where the physiological appropriateness of a generic modification can be demonstrated. It is therefore not subject to the difficulty under which the advocates of natural selection labor, when necessitated to explain a structure as being a step in the advance towards, or in the recession from, any *unknown* modification needful to the existence of the species. In the present case observation on the species in a state of nature may furnish interesting results. In no specimen has a trace of anything representing the lens been found."

When we remember that the lens of the eye in Amblyopsis has been found, even though the eye is less developed in all its parts than in Gronias, it is probable that a careful microscopical examination would show its existence in this genus also.

It is interesting to note that this fish is black above (lighter on the sides and white below), notwithstanding its supposed subterranean habits, and that all the other members of the family having rudimentary or covered eyes are also dark colored, while the blind fishes of the Mammoth Cave and of the caves in Cuba are nearly colorless. This want of color in the latter fishes has been considered as due to their subterranean life. If this be the cause, why should the blind cat fishes retain the colors characteristic of the other members of the family living in open waters?

* Proceedings of the Academy of Natural Sciences of Philadelphia for 1864, p. 231.

The fishes which in a general way, so far as blindness, tactile sense and mode of life are concerned, come the nearest to the blind fishes of the Mammoth Cave, are those described by Prof. Poey* under the names of *Lucifuga subterraneus* and *L. dentatus*.† These fishes having the broad, flattened, fleshy head, with minute cilia, without external eyes, and inhabiting caves so similar in structure to the Mammoth Cave, make a comparison of them with the fishes of the Mammoth Cave most interesting. This is greatly enhanced by the fact that the Cuban fishes belong to a family of essentially marine habit, quite far removed from Amblyopsis. The fresh water ling (Lota), belonging to the same great group of fishes (though to a distinct family or subfamily) containing the cod on the one hand and the Cuban blind fish on the other, is probably the nearest fresh water relative of the Cuban fish, but

Fig. 1.

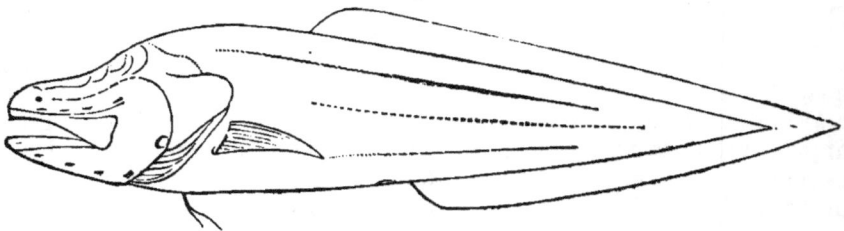

Blind Fish (*Stygicola dentatus*) from Caves in Cuba.

the nearest representative yet known is the marine genus Brotula, one species of which is found in the Caribbean Sea.

In the Cuban blind fish we find ciliary appendages on the head and body quite distinctly developed, evidently of the same character as those of Amblyopsis and answering the purpose of tactile organs. These cilia are in the form of small, but plainly visible, protuberances (reminding one of the single fleshy protuberance over the opercular opening just back of the head in Amblyopsis). There are eight of these on top of the head of a specimen I hastily examined, received from Prof. Poey by the Museum of Comparative Zoology, and quite a number arranged in three

---

* Memorias Sobre la Historia Natural de la Isla de Cuba. por Felipe Poey. Tomo 2, pp. 95–114. Pls. 9, 10, 11. Habana, 1856–8.

† This species was afterwards referred to the genus *Stygicola* Gill, on account of the presence of palatine teeth which are wanting in the other species. There are also several other good characters, to judge from the figures of the head, skull and brain given by Poey, that would warrant the reference of the fish to a distinct genus from *L. subterraneus*.

rows on each side of the body, showing that tactile sense is well developed in this fish ; though it is rather singular that the barbels on the jaws, so usually developed as organs of touch in the cod family and its allies, are entirely wanting in this fish.

The brain of *Lucifuga subterraneus*, as represented by the figures of Poey, differs very much from that of *L. dentatus* and of Amblyopsis. In all, the optic lobes are as largely developed as in allied fishes provided with well developed eyes. In *Lucifuga subterraneus* the cerebral lobes are separated by quite a space from the round optic lobes, which are represented as a little larger than the cerebral lobes, and also of greater diameter than the cerebellum ; this latter being more developed laterally than in either *L. dentatus* or in Amblyopsis. The three divisions of the brain are represented, from a top view, as nearly complete circles (without division into right and left lobes), of which that representing the optic lobes is slightly the largest. In *L. dentatus* the procencephalon and the optic lobes are represented as divided into right and left lobes, as in Amblyopsis, and the cerebellum does not extend laterally over the medulla oblongata as in *L. subterraneus*, but, as in Amblyopsis (Pl. 1, fig. 1 *d*), is not so broad as the medulla, and, projecting forwards, covers a much larger portion of the optic lobes than is the case in *L. subterraneus*.

The Cuban blind fish has the body, cheeks and opercular bones covered with scales. As in Amblyopsis the eyes exist, but are so imbedded in the flesh of the head as to be of no use. The outline cut here given (Fig. 1), copied from Poey, is very characteristic of the form of the fish, but does not exhibit the fleshy cilia or details of scaling.

The first notice that I can find of the Mammoth Cave blind fish is that contained in the " Proceedings of the Academy of Natural Sciences of Philadelphia," Vol. 1, page 175, where is recorded the presentation of a specimen to the Academy by W. T. Craige, M. D., at the Meeting held on May 24, 1842, in the following words :—

" A white, eyeless crayfish (*Astacus Bartoni?*) and a small white fish, also eyeless (presumed to belong to a subgenus of Silurus), both taken from a small stream called the 'River Styx' in the Mammoth Cave, Kentucky, about two and one-half miles from the entrance."

Dr. DeKay in his "Natural History of New York, Fishes," page 187, published in 1842, describes the fish, from a poor specimen in the Cabinet of the Lyceum of Natural History of New York, under the name of *Amblyopsis* * *spelæus*.† DeKay's description is on the whole so characteristic of the fish as to leave no doubt as to the species he had before him, though the statement that it has eight rays supporting the branchiostegal membrane (instead of six), and that the eyes are "large" but under the skin, must have been due to the bad condition of his specimen and to his taking the fatty layer covering the minute eyes for the eyes themselves, as pointed out by Prof. Wyman. Dr. DeKay places the genus with the Siluridæ (cat fishes) but at the same time questions its connection with the family and says that it will probably form the type of a new family. In 1843 Prof. Jeffries Wyman‡ gave an account of the dissection of a specimen in which he could not find a trace of the eye or of the optic nerve, probably owing to the condition of the specimen, as he afterwards§ found the eye spots, and made out the structure of the eye. When describing the brain, Prof. Wyman calls attention to the fact of the optic lobes being as well developed as in allied fishes with well developed eyes, and asks if this fact does not indicate that the optic lobes are the seat of other functions as well as that of sight. He also calls attention to the papillæ on the head as tactile organs furnished with nerves from the fifth pair.

Dr. Theo. Tellkampf‖ was the first to point out the existence of the rudimentary eyes from dissections made by himself and Prof. J. Müller, and to state that they can be detected in some specimens as black spots under the skin by means of a powerful lens. Prof. Wyman afterwards detected the eye through the skin in several specimens. Dr. Tellkampf also was the first to call attention to the " folds on the head, as undoubtedly serving as organs of touch, as numerous fine nerves lead from the trigeminal nerve to them and to the skin of the head generally."

It is also to Dr. Tellkampf that we are indebted for the first figure of the fish,¶ and for figures illustrating the brain, and internal organs. The descriptions of the anatomy of the fish by Drs.

* Obtuse vision.   † Of a cave.
‡ Silliman's Journal, Vol. 45, p. 94.
§ Proceedings Boston Soc. Nat. Hist.. Vol. 4. p. 395.   1853.
‖ Müller's Archiv. fur Anat., 1844. p. 392. Reprinted in the New York Journal of Medicine for July, 1845. p. 84, with plate.
¶ The only other figures of the species, that I am aware of, are the simple outlines given in Poey's Mem. de Cuba, the woodcut in Wood's Illustrated Natural History and the cut in Tenney's Zoology. None of these figures are very satisfactory.

Tellkampf and Wyman are all that have ever been written on the subject of any importance, with the exception of the description of the eye by Dr. Dalton, whose paper, in the "New York Medical Times," vol. 2, p. 354, I have not seen. Prof. Poey gives a comparison of portions of the structure with that of the Cuban blind fishes.

Dr. Tellkampf proposed the name of *Heteropygii** for the family of which, at the time, a single species from the Mammoth Cave was the only known representative, and makes a comparison of the characters with those of *Aphredoderus Sayanus*, a fish found only in the fresh waters of the United States, and belonging to the old family of Percoids, but now considered as representing a family by itself, though closely allied to the North American breams (Pomotis), and having the anal opening under the throat as in the blind fish.

Dr. Storer,† not knowing of Dr. Tellkampf's paper, proposed the name of *Hypsœidœ*, for the blind fish, and placed it between the minnow and the pickerel families, in the order of Malacopterygian, or soft rayed, fishes. According to the system adopted by Dr. Günther, it stands as closely allied to the minnows, *Cyprinodontidœ* (many of which are viviparous and have the single ovary and general character of the blind fish), and the shiners, *Cyprinidœ*, of the order of Physostomi. Dr. Tellkampf, in discussing the relations of the family, points out its many resemblances to the family of Clupesoces, and its differences from the Siluroids, Cyprinodontes and Clupeoids, with which it has more or less affinity, real or supposed. Prof. Cope in his paper on the Classification of Fishes ‡ places the Amblyopsis in the order of Haplomi with the shore minnows, pickerel and mud fish, and in an article on the Wyandotte Cave,§ he says that the Cyprinodontes (shore minnows) are its nearest allies. This arrangement by Prof. Cope places the Haplomi between the order containing the herrings and that containing the electric eel of South America, all included with the garpike, dog fish of the fresh waters (Amia), cat fishes, suckers and eels proper, etc., etc., in the division of Physostomi as limited by him.

---

*From the advanced position of the terminus of the intestine being so different from the position which it has in ordinary fishes.

† Synopsis of the Fishes of North America, published in 1846.

‡ American Naturalist, Vol. 5, p. 579, 1871.

§ Indianapolis Daily Journal of September 5, 1871. Reprinted in Ann. Mag. Nat. Hist., Nov., 1871.

Prof. Agassiz in 1851 [*] stated that the blind fish was an aberrant form of the Cyprinodontes.

Thus all those authors who have expressed an opinion as to the position which the fish should hold in the natural system have come to the same conclusions as to the great group, division, or order, into which it should be placed.   For all the terms used above, when reduced to any one system, bring Amblyopsis into the same general position in the system ; its nearest allies being the minnows, pickerels, shiners and herrings ; and unless a careful study of its skeleton should prove to the contrary, we must, from present data, consider the family containing Amblyopsis as more nearly allied to the Cyprinodontes, or our common minnows having teeth on the jaws, than to any other family, differing from them principally by the structure of the several parts of the alimentary canal and the forward position of its termination.

I have thus far mentioned only one species of blind fish from the cave, the *Amblyopsis spelæus*. The waters of the cave not only contain another species of blind fish, differing from Amblyopsis in several particulars, especially by its smaller size and by being without ventral fins, which I have identified as the *Typhlichthys subterraneus* of Dr. Girard ; but also a fish with well developed eyes, as proved by the account given by Dr. Tellkampf and by the drawing of a fish found by Prof. Wyman, in 1856, in the stomach of an Amblyopsis he was dissecting. In order to call attention to the fact that fishes with eyes are at times, if not always, in the waters of the cave, I have reproduced the drawing by Prof. Wyman on plate 1, fig. 13. It is very much to be regretted that the specimen is not now to be found, and that it was so much acted on by the gastric juice as to destroy all external characters by which it could be identified from the drawing, which is of about natural size. Dr. Tellkampf's remarks on the fish with eyes are as follows : —

"Besides the colorless blind-fish, there are also others found in the cave, which are black, commonly known by the name of 'mud-fish.' I saw a dark-colored fish in the water, but did not succeed in catching it. The latter are said to have eyes, and are entirely dissimilar to the blind-fish."

The name "mud-fish," given to this fish with eyes, and the statement that it is of a dark color, together with the drawing by Prof.

Wyman of the fish found in the stomach of the blind fish, showing the position of the dorsal fin to be the same as in the fish commonly called mud fish in the fresh waters of the Middle, Western and Southern States, perhaps, indicates the fish with eyes to be a species of *Melanura*.* This fish is called mud fish from the habit it has of burying itself in the mud, tail first, † to the depth of two to four inches, and of remaining buried in the mud in our western ditches during a time of drought. This habit, perhaps, in a measure fits it for a subterranean life. The occurrence of a fish belonging to the same family with the blind fish, but with well developed eyes, in the subterranean streams in Alabama, as mentioned further on and figured on Pl. 2, fig. 4, however, renders it probable that the cave fish with eyes may be the same or an allied species, and the drawing by Prof. Wyman would answer equally as well for it.

The fact that the Amblyopsis succeeded in catching a fish of, probably, very rapid and darting movements, shows that the tactile sense is well developed and that the blind fish must be very active in the pursuit of its prey; probably guided by the movement which the latter makes in the water so sensibly influencing the delicate tactile organs of the blind fish that it is enabled to follow rapidly, while the pursued, not having the sense of touch so fully developed, is constantly encountering obstacles in the darkness.

In describing the habits of the blind fish Dr. Tellkampf says :—

"It is found solitary, and is very difficult to be caught, since it requires the greatest caution to bring the net beneath them without driving them away. At the slightest motion of the water they dart off a short distance and usually stop. Then is the time to follow them rapidly with a net and lift them out of water. They are mostly found near stones or rocks which lie upon the bottom, but seldom near the surface of the water."

Prof. Cope, in describing the habits of the blind fish which he

---

* Dr. Günther considers the genus Melanura of this country to be synonymous with Umbla of Europe. In each country only one species has been as yet satisfactorily described.

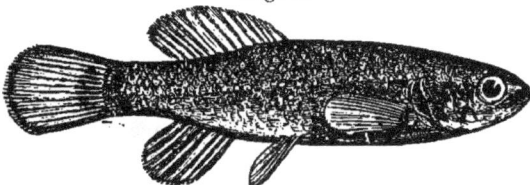

Fig. 2.

Mud fish (*Melanura limi*).

† See the interesting notes on the habits of the mud minnow, by Dr. Abbott in American Naturalist, Vol. 4, pages 107 and 388, with figure of the fish on page 385, which we here reproduce for comparison.

obtained in a stream that passes into the Wyandotte Cave, though he entered it by means of a well in the vicinity of the cave, says that : —

"If these Amblyopses be not alarmed they come to the surface to feed, and swim in full sight like white aquatic ghosts. They are then easily taken by the hand or net, if perfect silence be preserved, for they are unconscious of the presence of an enemy except through the sense of hearing. This sense is, however, evidently very acute, for at any noise they turn suddenly downward, and hide beneath stones, etc., on the bottom. They must take much of their food near the surface, as the life of the depths is apparently very sparse. This habit is rendered easy by the structure of the fish, for the mouth is directed upwards, and the head is very flat above, thus allowing the mouth to be at the surface."

The blind fish has a single ovary, in common with several genera of viviparous Cyprinodontes. In three female specimens of Amblyopsis which I have opened, the ovary was distended with large eggs, but no signs of the embryo could be traced. In these three specimens it was the right ovary that was developed, and this, as in the figure (Plate 2, fig. 1 c), was by the side of the stomach and did not extend beyond it. The number of eggs contained in the ovary was not far from one hundred in the specimen figured. As the embryos develop, the mass probably pushes further back in the cavity and also extends the abdominal walls. That the fish is viviparous is proved by the statement made by Mr. Thompson before the Belfast Natural History Society,* that one of the blind fishes from the cave, four and a half inches long, "was put in water as soon as captured, where it gave birth to nearly twenty young, which swam about for some time, but soon died. These, with the exception of one or two, were carefully preserved, and fifteen of them are now before us [at the meeting. I wish they were here], they were each four lines in length."

It is singular that no mention is made regarding these young, as to the presence or absence of eyes, and, as if it was fated that this important point should remain unnoticed as long as possible, it is equally singular that Dr. Steindachner omitted to examine some very young specimens which he received from a friend a few months since and sent to the Vienna Museum, where they will remain unexamined until he returns there. I saw the Doctor only

---

* Annals and Mag. of Natural History, Vol. xiii, pp. 112, 1844.

a week after these, to me, interesting specimens had been sent abroad, and he was as grieved as I was disappointed at my being just too late to take advantage of them. (See note on p. 52.)

At what time the young are born has never been stated, but judging from such data as I can at present command, I think that it must be during the months of September and October. Specimens collected during those months would probably contain embryos in various stages of development, the examination of which would undoubtedly lead to most interesting results. (See note on p. 52.)

Prof. Wyman has most generously placed in my hands his unpublished notes and drawings of the several dissections he has made of Amblyopsis, as well as his specimens and dissections. Many of these drawings are reproduced on Plate 1, and will, with his notes which I here give, greatly enhance the value of this article, as his dissections have been made with the utmost care, and with a patience and delicacy that only a master hand attains. It will therefore be understood that, in giving credit to Prof. Wyman in the following pages, I refer to his unpublished notes, except when the quotation is given from a special work. In quoting his description of the eye and ear from "Silliman's Journal" I have changed the references so as to refer to his drawings reproduced on Plate 1, and not to the three cuts given in "Silliman's Journal," though the figures of the brain and of the otolite were copied from those cuts.

The largest specimens I have seen of Amblyopsis are several males and females, each from four to four and a half inches in length, which seems to be about as large as the fish grows, though Dr. Günther mentions a specimen in the British Museum of five inches in length. The largest specimen captured of late years is said to have been taken, during the summer of 1871, and sold for ten dollars to a person who was so desirous of securing the precious morsel that he had it cooked for his supper. The smallest specimen I have seen was one and nine-tenths inches in length. The general shape and character of the fish is best shown by the figures on plates 1 and 2.

"The whole head, above and below, is destitute of scales, the naked skin extending backwards on the sides to the base of the pectoral fins ; the scaly portion of the body above ends in a semicircular edge covering the space between the upper ends of the opercula. The skin covering the middle region of the head is smooth, but on either side is provided with numerous transverse and longitudinal ridges (Pl. 1, fig. 7), which are, on the whole, regularly arranged. The first row of transverse ridges, eight or nine in number, begins between the nostrils and extends backwards, diverging from the median line. The third ridge is crossed at its outer end by a longitudinal one, as are also two others farther back. The second and third rows, situated, in part, on the sides and, in part, on the under surface, are less regular than the preceding. A fourth, on the borders of the operculum, is still less well defined. The transverse are also crossed here by longitudinal ridges. About ten vertical ridges, also provided with papillæ, and similar to those on the head, are visible on the sides extending from the pectoral fins to the tail, but are not so well defined as those on the head. The skin of the head is of extreme delicacy and is covered by a very thin, loose layer of epithelium."—WYMAN.

"The larger ridges have between twenty and thirty papillæ, many of these having a cup-shaped indentation at the top, in which a delicate filament is, in some instances, seen (Pl. 1, fig. 9). These papillæ are largely provided with nervous filaments, and, as is obvious from their connection with branches of the fifth pair of nerves, must be considered purely tactile, and the large number of them shows that tactile sensibility is probably very acute and in some measure compensates for the virtual absence of the sense of sight. Plate 1, fig. 8, represents one of the ridges of the head magnified, showing the papillæ of which it is made up, and figure 9 shows three papillæ still more enlarged. Two of these show a cup-shaped cavity at the top, and the short, slender filament already mentioned. The surface of the papillæ is covered with loosely connected epithelium cells. Fig. 10 shows the nervous filaments distributed to the papillæ: a, a branch of the fifth pair of nerves passing beneath the papillary ridge and sending filaments to each papilla. These papillary branches interchange filaments, forming a nervous plexus in connection with each ridge. This figure of the nerves was drawn with a camera lucida, from a specimen treated with acetic acid."— WYMAN.

"Plate 1, fig. 6, represents a double system of subcutaneous canals, which extend the whole length of the head, but were not traced farther back than the edge of the naked or scaleless skin which covers it. Forwards they bifurcate, nearly encircling the nasal cavity, towards the middle line ending in a blind pouch.

The lateral branch was not traced distinctly to an end, but seemed to connect with the olfactory cavity. The walls of these canals are exceedingly delicate and easily overlooked."—WYMAN.

"Plate 1, fig. 5, shows the globe of the eye with the optic nerve (c), as seen under the microscope. The lens (b) is detached from its proper place by the pressure of the glass. Irregularly arranged muscular bands are attached to the exterior of the globe (a, a, a, a), but were not recognized as the homologues of the muscles of the normal eye of fishes ; nevertheless, they indicate that the globe was movable."—WYMAN.

"In the three specimens recently dissected, the eyes were exposed only after the removal of the skin, and the careful separation from them of the loose areolar tissue which fills the orbit.   In a fish four inches in length the eyes measured one-sixteenth of an inch in their long diameter, and were of an oval form and black.   A filament of nerve (Pl. 1, fig. 3 a) was distinctly traced from the globe to the cranial walls, but the condition of the contents of the cranium, from the effects of the alcohol, was such as to render it impracticable to ascertain the mode of connection of the optic nerve with the optic lobes.

Examined under the microscope with a power of about twenty diameters, the following parts were satisfactorily made out (Pl. 1, fig. 3) : 1st, externally an exceedingly thin membrane, b, which invested the whole surface of the eye and appeared to be continuous with a thin membrane covering the optic nerve, and was therefore regarded as a sclerotic ; 2d, a layer of pigment cells, d, for the most part of a hexagonal form, and which were most abundant about the anterior part of the eye ; 3d, beneath the pigment a single layer of colorless cells, c, larger than a pigment cell, and each cell having a distinct nucleus ; 4th, just in front of the globe ; a lenticular-shaped, transparent body, e [see also fig. 4], which consisted of an external membrane containing numerous cells with nuclei.   This lens-shaped body seemed to be retained in its place by a prolongation forwards of the external membrane of the globe ; 5th, the globe was invested by loose areolar tissue, which adhered to it very generally, and in some instances contained yellow fatty matter ; in one specimen it formed a round spot, visible through the skin on each side of the head, which had all the appearance of a small eye ; its true nature was determined by the microscope only.   It is not improbable that the appearance just referred to may have misled Dr. DeKay—where he states that the eye exists of the usual size, but covered by the skin.

If the superficial membrane above noticed is denominated correctly the *sclerotic*, then the pigment layer may be regarded as the representation of the *choroid*.   The form as well as the position of the transparent nucleated cells within the choroid correspond for the most part with the *retina*. All of the parts just enumerated are such as are ordinarily developed from and in connection with the encephalon, and are not in any way dependent upon the skin. But if the lenticular-shaped body is the true representative of the crystalline lens, it becomes difficult to account for its presence in Amblyopsis according to the generally recognized mode of its development (since it is usually formed from an involution of the skin) unless we suppose that after the folding in of the skin had taken place in the embryonic condition, the lens retreated from the surface, and all connection with the integument ceased.[*]

According to Quatrefages, however, the eye of Amphioxus [†] is contained wholly in the cavity of the dura mater, and yet it has all the appearance of being provided with a lens. If his description be correct, then the mode of development as well as the morphology of the eye in this remarkable fish is different from that of most other vertebrates, since the lens never could have been formed from an involution of the skin, nor could the eye with its lens, as Prof. Owen asserts, be a modified cutaneous follicle. Whatever views be taken with regard to the development of the eye of the blind fish, the anatomical characters which have been enumerated show, that though quite imperfect as we see it in the adult, it is constructed after the type of the eyes of other vertebrates. It certainly is not adapted to the formation of images, since the common integument and the areolar tissue which are interposed between it and the surface, would prevent the transmission of light to it except in a diffused condition. No pupil or anything analogous to an iris was detected, unless we regard as representing the latter the increased number of pigment cells at the anterior part of the globe.

It is said that the blind fishes are acutely sensitive to sounds as well as to undulations produced by other causes in the water. In the only instance in which I have dissected the organ of hearing (which I believe has not before been noticed), all its parts were largely developed, as will be seen by reference to Pl. 1, fig. 1 *e*. As regards the general structure, the parts do not differ materially from those of other fishes except in their proportional dimensions. The semi-circular canals are of great length, and the two which unite to enter the vestibule by a common duct, it will be seen, project upwards and inwards under the vault of the cranium, so as to approach quite near to the corresponding parts of the opposite side. The otolite contained in the utricle was not remarkable, but that of the vestibule (Pl. 1, fig. 2) and seen in

---

[*] In birds and mammals there is a stage of development where the lids come together and firmly unite, to separate again when the animal "gets its eyes open." In the mole rat (*Spalax typhlus*) of Siberia, the lids never open, and the eyes remain through life covered with hairy skin. It is not improbable that in Amblyopsis something analogous to this, a closing of the skin over the eye, may have taken place.—J. W.

[†] I have used the prior name of *Branchiostoma* in this paper when speaking of the Lancelet.

dotted outline in fig. 1 *e* is quite large when compared with that of a *Leuciscus* of about the same dimensions as the blind-fish here described."—WYMAN, *Silliman's Journal*, Vol. 17, *p*. 259, 1854.

The *Amblyopsis spelæus* undoubtedly has quite an extensive distribution, probably existing in all the subterranean rivers that flow through the great limestone region underlying the Carboniferous rocks in the central portion of the United States. Prof. Cope obtained specimens from the Wyandotte Cave and from wells in its vicinity, and in the Museum of Comparative Zoology at Cambridge there is a specimen labelled "from a well near Lost River, Orange Co., Ind.," which, with those from the Wyandotte Cave, is conclusive evidence of its being found on the northern side of the Ohio* as well as on the southern, in the rivers of the Mammoth Cave. I have been able to examine a number of specimens from the Mammoth Cave, and have carefully compared with them the one from the well in Orange Co., Ind., and find that the specific characters are remarkably constant.

In 1859 † Dr. Girard described a blind fish, received by the Smithsonian Institution from J. E. Youngglove, Esq., who obtained it "from a well near Bowling Green, Ky." The general appearance of this fish, which was only one and a half inches in length, was that of *Amblyopsis spelæus*, but it differed from that species in several characters, especially by the *absence* of ventral fins. Dr. Girard therefore referred the fish to a distinct genus under the name of *Typhlichthys‡ subterraneus*. Dr. Günther § considers this fish a variety of *Amblyopsis spelæus* and records the specimen in the British Museum "from the Mammoth Cave," as "half-grown." ‖

By the kindness of Prof. Agassiz, I have been enabled to examine nine specimens of *blind fish without ventrals*, in the Museum of Comparative Zoology. Seven of these were collected in the Mammoth Cave by Mr. Alpheus Hyatt in September, 1859. One was from Moulton, Lawrence County, Alabama, presented by Mr. Thomas Peters; and another from Lebanon, Wilson Co., Tennessee; presented by Mr. J. M. Safford. It is not stated whether

---

* I have also been informed by Mr. Holmes of Lansing, Mich., that *blind fishes* have been drawn out of wells in Michigan.
† Proceedings Acad. Nat. Sci. Philad., p. 63.
‡ Blind fish.
§ Catalogue of Fishes in the British Museum, Vol. 7, p. 2, 1868.
‖ The largest specimen I have seen of Typhlichthys is one and seventeen-twentieths inches in length, and the smallest Amblyopsis one and eighteen-twentieths inches.

these latter came from wells or caves, but probably from wells.
They are all of about the same size, one and one-half to two
inches in length, and are constant in their characters. Moreover,
four of the seven specimens from the Mammoth Cave were females
with eggs. These eggs were as large in proportion as those from
Amblyopsis. The ovary was single and situated on the right
side of the stomach, as in Amblyopsis. The difference in the
number of eggs was very remarkable, each of the four specimens
examined having but about thirty eggs in the ovary, while in
three females of Amblyopsis (all, however, of nearly three times
the size of Typhlichthys) there were about one hundred eggs in
each. As in both species there were no signs of the embryos in
the eggs, it is not probable that any of the eggs had been developed
and the young excluded, nor is it at all likely that the great variation in the number of eggs would simply indicate different ages.
By a reference to the figures (Pl. 2), it will be seen that the pyloric
appendages, stomach and scales of the two fishes are different.
For these reasons, taken in connection with the absence of ventral fins, I have no hesitation in accepting Dr. Girard's name as
valid for this genus, of which we thus far know of but one species,
with a subterranean range from the waters of the Mammoth Cave,
south to the northern portion of Alabama. In this connection it
would be most interesting to know the relations of the "blind
fishes" said to have been found in Michigan. For thus far we
have Typhlichthys limited to the central and southern portion of
the subterranean region, Amblyopsis to the central, and the species in the northern portion undetermined.

In 1853, on his return from a tour through the southern and
western states, Prof. Agassiz gave a summary of some of his
ichthyological discoveries in a letter to Prof. J. D. Dana.* In this
letter are the following remarks : —

"I would mention foremost a new genus which I shall call *Chologaster*, very similar in general appearance to the blind fish of the
Mammoth Cave, though provided with eyes; it has, like Amblyopsis, the anal aperture far advanced under the throat, but is entirely deprived of ventral fins; a very strange and unexpected
combination of characters. I know but one species, *Ch. cornutus*
Ag. It is a small fish scarcely three inches long, living in the
ditches of the rice fields in South Carolina. I derive its specific

name from the singular form of the snout, which has two horn-like projections above."

This is the only information ever published regarding this interesting fish and the only specimens known are those on which Prof. Agassiz based the above remarks.

By the kindness of Professor L. Agassiz, who has placed all the specimens of the family contained in the Museum of Comparative Zoology in my hands for study, I am enabled to give a figure and description of this interesting species from the three specimens in the Museum, which were labelled as the originals of *Chologaster cornutus* Ag., from Waccamaw, S. C., presented by Mr. P. C. J. Weston, 1853. The largest of these specimens was distended with eggs and I was enabled to compare the ovary with that of Amblyopsis. From its being single and the eggs very large, I have no doubt that it is a viviparous fish like the other genera of the family. The position of the ovary behind the stomach, as shown in fig. 2 c, plate 2, and the presence of four pyloric appendages (Pl. 2, fig. 2 a) instead of two, as in Amblyopsis (fig. 1 a) and Typhlichthys (fig. 3 a), are good internal characters, separating it from the other genera, independently of the presence of eyes and the absence of ventral fins and papillary ridges.

The stability of the internal characters I have mentioned was most unexpectedly substantiated by the discovery of a second species (Pl. 2, figs. 4, 4 a) of the genus among the specimens in the Museum of Comparative Zoology. I have the pleasure of dedicating this species* to Professor Agassiz, not only in kindly remembrance of the eight years I was associated with him as student and assistant, but also because the fish so well illustrates the decided position he has taken relative to the immutability of species.

The only specimen known of this second species was drawn from a well in Lebanon, Tenn., and presented to the Museum by Mr. J. M. Safford, Jan., 1854. It is a more slender fish than *C. cornutus*, but the intestine follows the same course and the four pyloric appendages are present as in that species.

In the genus Chologaster† we have all the family characters as well expressed as in the blind species, though it differs from Am-

---

* A Synopsis of this family with descriptions of the four species will appear in the " Report of the Peabody Academy of Science for 1871." (Reprinted here. p. 55.)

† Literally " bile-stomach;" probably named from the yellow color of the fish.

blyopsis and Typhlichthys by the presence of eyes, the absence of papillary ridges on the head and body, and by the longer intestine and double the number of pyloric appendages, as well as by the position of the ovary ; and agrees with Typhlichthys by the absence of ventral fins. Amblyopsis and Typhlichthys are nearly colorless, while *Chologaster Agassizii* is of a brownish color similar to many of the minnows, and *C. cornutus* is brownish yellow, with dark, longitudinal bands.

Among the most interesting points in the history of this genus is the fact of its occurring in two widely different localities, *C. Agassizii* having been found in a well, in the same vicinity (probably in the same well) with a specimen of Typhlichthys, and undoubtedly belonging to the same subterranean fauna west of the Appalachian ridge, while *C. cornutus* belongs to the southern coast fauna of the eastern side of that mountain chain, and is thus far the only species of the family known beyond the limits of the great subterranean region of the United States.

Having now given an outline of the structure, habits and distribution of the four species belonging to the family, and recapitulated the known facts, we are better able to consider the bearings of the peculiar adaptation of the blind fishes, in the Mammoth and other caves, to the circumstances under which they exist.

Prof. Cope in stating, in his account of the blind fish of the Wyandotte Cave, " that the projecting under jaw and upward direction of the mouth renders it easy for the fish to feed at the surface of the water, where it must obtain much of its food," suggests that : —

" This structure also probably explains the fact of its being the sole representative of the fishes in subterranean waters. No doubt many other forms were carried into the caverns since the waters first found their way there, but most of them were like those of our present rivers, deep water or bottom feeders. Such fishes would starve in a cave river, where much of the food is carried to them on the surface of the stream. . . . . . The shore minnows are their nearest allies, and many of them have the upturned mouth and flat head. . . . . Fishes of this, or a similar family, enclosed in subterranean waters ages ago, would be more likely to live than those of the other, and the darkness would be very apt to be the cause of the atrophy of the organs of sight seen in the Amblyopsis."

This suggestion was undoubtedly hastily made by Prof. Cope when writing the letter which was printed in the " Indianapolis

Journal," and were it not that the article has been reprinted in the "Annals and Magazine of Natural History," I should not criticise the statement made in an off-hand letter for publication in a newspaper; for with Prof. Cope's knowledge of fishes it could simply be a hasty thought which he put on paper, when he suggests that it is because the Cyprinodontes have a mouth directed upwards and are surface feeders that they were better adapted to a subterranean life than other fishes, and hence maintained an existence, while other species, which he supposes were introduced into the subterranean streams at the same time, died out.

If the fishes of the subterranean streams came from adjoining rivers, why were not many of the Percoids, Cyprinoids and other forms, that are as essentially surface feeders as the Cyprinodontes (many of the latter are purely "mud feeders"), as capable of maintaining an existence in the subterranean waters as any species of the latter? Neither is it necessary for us to assume that the structure of the fish should be adapted to feeding on the surface, for not only have we in the blind cat fish, described by Prof. Cope himself, from the subterranean stream in Pennsylvania, an example of a fish belonging to an entirely different family of bottom feeders, thriving under subterranean conditions, but the blind fishes of the Cuban caves are of the great group of cod fishes which are, with hardly an exception, bottom feeders. The fact that the food of the blind fishes of the Mammoth Cave consists in great part of the cray fish found in the waters of the cave, as shown by the contents of several stomachs I have examined, and also that one blind fish at least made a good meal of another fish, as already mentioned, shows that they are not content with simply waiting for what is brought to them on the surface of the water, and that they are probably as much bottom as surface feeders.

Again, in regard to the sense of sight, why is it necessary to assume that because fishes are living in streams where there is little or no light, that it is the cause of the non development of the eye and the development of other parts and organs? If this be the cause, how is it that the Chologaster from the well in Tennessee, or the "mud fish" of the Mammoth Cave are found with eyes? Why should not the same cause make them blind if it made the Amblyopsis and Typhlichthys blind? Is not the fact, pointed out by Prof. Wyman, that the optic lobes are as well developed in Amblyopsis as in allied fishes with perfect eyes, and, I may add, as well developed as those of *Chologaster cornutus*, an argument in favor of the theory that the fishes were always blind and that they have not become so from the circumstances under which they exist? If the latter were the case and the fishes have become blind from the want of use of the eyes, why are not the optic lobes also atrophied, as is known to be the case when other animals lose their sight? I know that many will answer at once that Amblyopsis and Typhlichthys have gone on further in the development and retardation of the characters best adapting them to their subterranean life, and that Chologaster is a very interesting transitionary form between the open water Cyprinodontes and the subterranean blind fishes. But is not this assumption answered by the fact that Chologaster has every character necessary to place it in the same family with Amblyopsis and Typhlichthys, while it is as distinctly and widely removed from the Cyprinodontes as are the two blind genera mentioned?

Assuming, for the moment, that Chologaster is a transitional form between the surface feeding Cyprinodontes, and Typhlichthys and Amblyopsis, let us recapitulate the characters that distinguish the different forms and see if they exhibit transitions, and if Chologaster is traversing the slow developmental road to Amblyopsis.

Allowing all characters embraced in the general structure of the skeleton, brain, scales, fins, etc., as ordinal, and common to both Cyprinodontes and Heteropygii, we will recapitulate only such as can be considered of family and generic value in the two groups.

	CYPRINODONTES.	CHOLOGASTER.	TYPHLICHTHYS.	AMBLYOPSIS.
Surface feeders.	In part.	Unknown.	Partially.	The same.
Intestine.	In many genera long and convoluted, in others short and with single turn.	Moderately long with two turns.	Shorter with two turns.	The same.
Stomach & pyloric appendages.	In most, if not all, stomach not well defined from intestine and without appendages.	Stomach well defined, cœcal, with two pyloric appendages on each side.	The same, with one pyloric appendage on each side.	The same.
Viviparous.	Many genera.	Probably.	Probably.	Undoubtedly.
Ovary.	Single in viviparous genera* and placed by the side of intestine in some and posterior in others.	Single and placed behind the stomach.	Single and placed at side of stomach.	The same.

---

* The ovary is also single in other genera of viviparous fishes belonging to distinct orders.

Anal opening.	In normal position.	Forward of pectorals.†	The same.	The same.
Air bladder.	Present in few genera.‡	Present.	The same.	The same.
Scales.	On body regularly imbricated and loosely attached.	Irregularly arranged, firmly attached by being covered in great part by the cuticle.	The same.	The same.
Head with scales or naked.§	With scales.	Naked.	The same.	The same.
Tactile papillæ ‖ on the head and body.	Absent.	Absent.	Very prominent as ridges on the head and sides of body.	The same.
Ventral fins.¶	Present in most genera, absent in at least two.	Absent.	Absent.	Present.
*Eyes.***	Well developed in all.	Well developed and normal.	Rudimentary †† and of no use.	The same.
Habitat.	Fresh water; brackish water; salt water.	Limestone water of subterranean rivers. Brackish water?	Limestone water of subterranean rivers.	The same.
Geographical range.	Nearly all parts of the world.	One species in subterranean streams of S. central portion of the U. S.; a 2d species in the So. Atl. coast fauna of U. S.	Central & southern portion of subterranean fauna of United States.	Central and N. central portion of same.

From this brief comparison of some of the prominent characters of the genera of the Heteropygii with the Cyprinodontes, their

† Aphredoderus and Gymnotus, and other genera of distinct orders have this forward position of the anus also.

‡ The air bladder is in several families present in some species and absent in others.

§ The presence or absence of scales on the head, or on portions of it, is a generic character subject to great variation in many families and quite constant in others.

‖ I cannot recall anything but the barbels on the head and jaws of many genera of Cyprinoids, Siluroids, Gadoids, etc., etc., that can be said to be tactile organs among fishes, with the exception of the fleshy papillæ on the head and body of the blind fishes of the American and Cuban caves, and the filaments of the fin rays of many fishes and the fleshy ventral rays of the Gurnards.

¶ Of all fins, the ventrals are the most likely to deviate from their normal structure and position. Their presence or absence, as exhibited in many families, and often by different ages of the same fish, and the great variation in their position in different genera of the same family, render any change in them of either generic, specific, or individual character, or simply indicative of age (as they are lost in some adult fishes while present in the young, and in others not developed until after the other fins).

** As I have alluded to the fact, in the first part of this paper, the eyes of fishes are no more the constant and unvarying part of the fish structure than the ventral fins, and like them are subject to almost every conceivable variation in position in the head, and perfection in structure.

†† The largest specimen I have seen of Typhlichthys, is less than two inches in length and as the eye of an Amblyopsis of twice the size is not over a 32d of an inch in width it must be very small indeed in Typhlichthys, and I confess to not being able to find it in an ordinary dissection, assisted only by a good lens.

acknowledged nearest allies, we can only trace what could be regarded as a transition, or an acceleration, or a retardation of development, in simply those very characters, of eyes and ventral fins, that are in themselves of the smallest importance in the structure (permanence of character considered) of a fish, and, as if to show that they were of no importance in this connection, we find in the same cave, blind fishes with ventrals and without; and in the same subterranean stream, a blind fish and another species of the family with well developed eyes.

If it is by acceleration and retardation of characters that the Heteropygii have been developed from the Cyprinodontes, we have indeed a most startling and sudden change of the nervous system. In all fishes the fifth pair of nerves send branches to the various parts of the head, but in the blind fishes these branches are developed in a most wonderful manner, while their subdivisions take new courses and are brought *through* the skin, and their free ends become protected by fleshy papillæ, so as to answer, by their delicate sense of touch, for the absence of sight. At the same time the principle of retardation must have been at work and checked the development of the optic nerve and the eye, while acceleration has caused other portions of the head to grow and cover over the retarded eye.

Now, if this was the mode by which blindness was brought about and tactile sense substituted, why is it that we still have *Chologaster Agassizii* in the same waters, living under the same conditions, but with no signs of any such change in its senses of sight and touch? It may be said that the Chologaster did not change because it probably had a chance to swim in open waters and therefore the eyes were of use and did not become atrophied. We can only answer, that if the Chologaster had a chance for open water, so did the Typhlichthys and yet that is blind.

If the Heteropygii have been developed from Cyprinodontes, how can we account for the whole intestinal canal becoming so singularly modified, and what is there in the difference of food or of life that would bring about the change in the intestine, stomach and pyloric appendages, existing between Chologaster and Typhlichthys in the same waters? To assume, that under the same conditions, one fish will change in all these parts and another remain intact, by the blind action of uncontrolled natural laws, is, to me, an assumption at variation with facts as I understand them.

Looking at the case from the standpoint which the facts force me to take, it seems to me far more in accordance with the laws of nature, as I interpret them, to go back to the time when the region now occupied by the subterranean streams, was a salt and brackish water estuary, inhabited by marine forms, including the brackish water forms of the Cyprinodontes and their allies (but not descendants) the Heteropygii. The families and genera having the characters they now exhibit, but most likely more numerously represented than now, as many probably became exterminated as the salt waters of the basin gradually became brackish and more limited, as the bottom of this basin was gradually elevated, and finally, as the waters became confined to still narrower limits and changed from salt to brackish and from brackish to fresh, only such species would continue as could survive the change, and they were of the minnow type represented by the Heteropygii, and perhaps some other genera of brackish water forms that have not yet been discovered.

In support of this hypothesis we have one species of the family. *Chologaster cornutus*, now living in the ditches of the rice fields of South Carolina, under very similar conditions to those under which others of the family may have lived in long preceding geological times ; and to prove that the development of the family was not brought about by the subterranean conditions under which some of the species now live, we have the one with eyes living with the one without, and the South Carolina species to show that a subterranean life is not essential to the development of the singular characters which the family possess.

That a salt or brackish water fish would be most likely to be the kind that would continue to exist in the subterranean streams. is probable from the fact that in all limestone formations caves are quite common, and would in most instances be occupied first with salt water and then brackish, and finally with fresh water so thoroughly impregnated with lime as to render it probable that brackish water species might easily adapt themselves to the change, while a pure fresh water species might not relish the solution of lime any more than the solution of salt, and we know how few fishes there are that can live for even an hour on being changed from fresh to salt, or salt to fresh, water. We have also the case of the Cuban blind fishes belonging to genera with their nearest representative in the family a marine form, and with the whole family of cods and their allies, to which group they belong, essentially marine. Further than this the cat fish from the subterranean stream in Pennsylvania belongs to a family having many marine and brackish water representatives. As another very interesting fact in favor of the theory that the Heteropygii were formerly of brackish water, we have the important discovery by Prof. Cope of the Lernæan parasite on a specimen of Amblyopsis from the Wyandotte cave ; this genus of parasitic crustaceans being very common on marine and migratory fishes, and much less abundant on fresh water species.

Thus I think that we have as good reasons for the belief in the immutability and early origin of the species of the family of Heteropygii, as we have for their mutability and late development, and, to one of my, perhaps, too deeply rooted ideas, a far more satisfactory theory ; for, with our present knowledge, it is but theory on either side.

YOUNG OF THE BLIND FISH. — Dr. Hagen gives me the following information about the young specimens I mentioned (page 38) as belonging to Dr. Steindachner, which I just missed seeing before they were sent to Vienna. These specimens were procured by Dr. Hartung for Dr. Steindachner under the following circumstances. Just as Dr. Hartung was leaving the cave hotel on Oct. 21, a bottle was brought to him containing four specimens, one of which was smaller than the others (probably Typhlichthys), all living. He immediately transferred them to a jar containing alcohol and took no notice of them until he reached Nashville, when he discovered an addition of *eight little ones* in the jar.

The birth of these young was undoubtedly due to placing the parent in the alcohol, and the date (Oct. 21) would correspond to the time I have stated as probably that at which the young were born.

Dr. Hagen states that he examined the young under a lens, without taking them from the jar, and *could not discover any eyes.* The specimens were about three lines in length.

So now we have two more facts to add to the history of the blind fishes (though whether they apply to Amblyopsis or Typhlichthys is not yet settled). First, that the young are born in October, and second, that they are without external eyes when born. — *From the* AMERICAN NATURALIST *for February,* 1872.

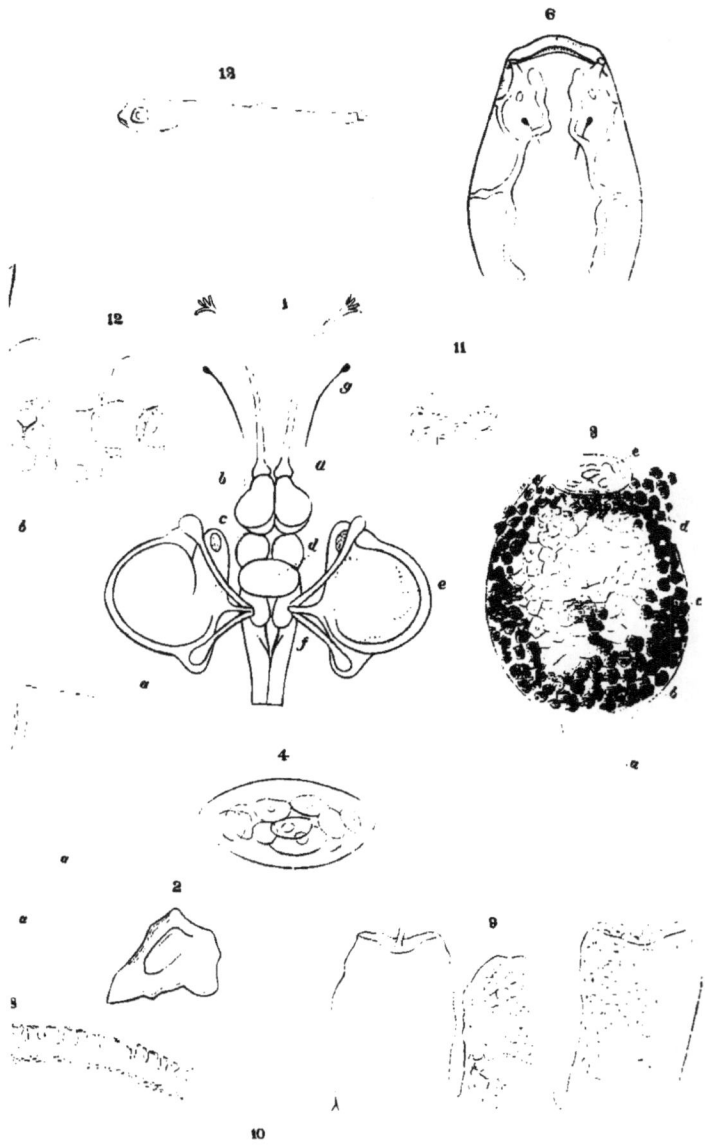

EXPLANATION OF PLATE ONE.

[All the figures on this plate are from original drawings by Prof. J. Wyman.]

FIG. 1. Brain, nerves and organ of hearing of *Amblyopsis spelæus;* enlarged; *a*, olfac tory lobes and nerves; *b*, cerebral lobes; *c*, optic lobes; *d*, cerebellum; *e*, organ of hearing, showing the semicircular canals, with the otolite represented in place by the dotted lines; *f*, medulla oblongata; *g*, optic nerves and eye specks.

FIG. 2. Otolite, enlarged.

FIG. 3. Eye, magnified (natural size one-sixteenth of an inch in length); *a*, optic nerve; *b*, sclerotic membrane; *c*, layer of colorless cells; *d*, layer of pigment cells (iris?); *e*, lens.

FIG. 4. Lens, enlarged and showing the cells.

FIG. 5. Eye, enlarged, showing the muscular bands, *a, a, a, a; b*, the lens pressed out of place; *c*, the optic nerve.

FIG. 6. Top of head, showing the canals under the skin, of the natural size. The two black dots and lines indicate the eyes and optic nerves in position.

FIG. 7. Top of head, showing the arrangement of the ridges of papillæ. Natural size.

FIG. 8. One of the ridges of papillæ from the head, magnified.

FIG. 9. Three of the papillæ from the ridge, still more magnified, showing the cup-shaped summit and projecting filament.

FIG. 10. A portion of the ridge magnified, and treated with acid, to show the arrangement of the nervous plexus supplying the papillæ with nerve filaments from a branch (*a*) of the fifth pair.

FIG. 11. Epithelial cells from the head.

FIG. 12. Epithelial cells from the body.

FIG. 13. A fish with eyes, found in the stomach of an Amblyopsis.

EXPLANATION OF PLATE TWO.

FIG. 1.   AMBLYOPSIS SPELÆUS DeKay. Natural size.
- 1*a*. Stomach and pyloric appendages. Twice natural size.
- 1*b*. Scale, magnified.
- 1*c*. Abdominal cavity, showing position of stomach and single ovary. Natural size.

FIG. 2.   CHOLOGASTER CORNUTUS Agassiz. Natural size.
- 2*a*. Stomach and pyloric appendages. Twice natural size.
- 2*b*. Scale, magnified.
- 2*c*. Abdominal cavity, showing stomach and single ovary behind the stomach. Twice natural size.

FIG. 3.   TYPHLICHTHYS SUBTERRANEUS Girard. Slightly more than natural size.
- 3*a*. Stomach and pyloric appendages. Twice natural size.
- 3*b*. Scale, magnified.

FIG. 4.   CHOLOGASTER AGASSIZII Putnam. Natural size.
- 4*a*. Stomach and pyloric appendages. Twice natural size.
- 4*b*. Scale, magnified.

The scales figured on the plate are all from the second or third row under the dorsal fin. 4*b* is represented with the posterior margin *down*, all the others are represented with the posterior margin on the *left*. The natural size of the scales is given by the minute outline at the left of the figures above each scale; 4*b* is so small that the natural size can hardly be represented by the black dot.

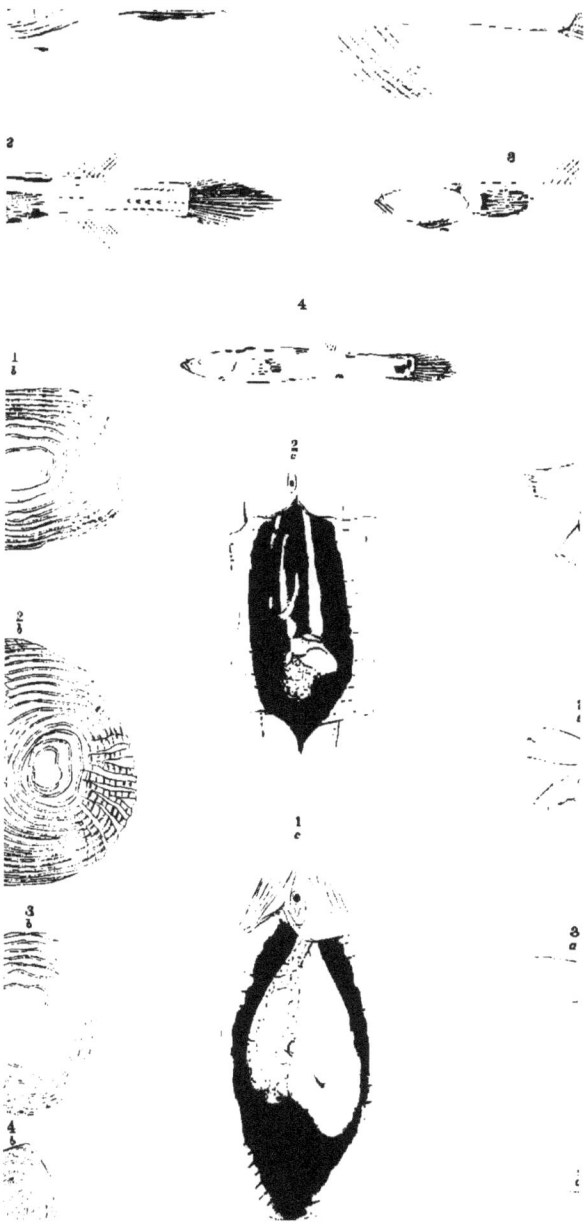

2

8

4

1
b

2
c

2
b

3
b

4
b

1
c

3
a

CHAPTER IV.

SYNOPSIS OF THE FAMILY HETEROPYGII.*

BY F. W. PUTNAM.

—•◦•—

HETEROPYGII Tellkampf, Müller's Arch. f. Anat., p. 392, 1844; and New York Journal of Medicine, v, p. 84, 1845.

Hypsæidæ Storer, Synopsis N. A. Fish, p. 435, 1846.

Brain of ordinary development in all its parts, similar to that of Cyprinodontes and of about the same proportions. Cerebral lobes larger than the nearly round optic lobes. Cerebellum overlapping the posterior third of the optic lobes. Medulla oblongata broad, with well defined right and left sides. (On comparing the brains of the three genera the only difference noticed was that in Chologaster the cerebellum was not quite as large proportionally, but more elongated and not quite as wide as in the other genera, while the optic lobes of this genus with well developed eyes were no larger than in a Typhlichthys of the same size.)

Skeleton not studied. Günther gives the vertebræ as thirteen abdominal and nineteen or twenty caudal. The bones of the head are thin and mostly flattened as in the Cyprinodontes. Occiput slightly convex.

Body compressed posteriorly. Head and anterior portion of body depressed, giving the form of a broad, flat head, with a compressed tail.

Branchiostegal rays six in number and but slightly covered by opercular bones; opercular opening large.

Fins. Dorsal and anal nearly opposite and posterior to centre of body. All the fins except the ventrals well developed, with central rays longest and first rays simple. Pectorals close to the head, about in the middle of the sides. (Ventrals present in Amblyopsis, absent in Typhlichthys and Chologaster.)

Mouth opening upwards, with lower jaw slightly projecting. Margin of the upper jaw formed by the intermaxillaries. Maxillaries placed behind the intermaxillaries, with lower third broad and below the intermaxillaries. Several rows of fine teeth on the intermaxillaries and lower jaw. (Teeth on palatines in Amblyopsis and Typhlichthys, none on these bones in adults of Chologaster.)

Scales. None on the head. Body closely covered with small, partially imbedded cycloid scales, irregularly arranged.

Lateral line absent.

---

* From the Annual Report of the Peabody Academy of Science for 1871.

Nostrils double. Anterior tubular and standing out from the end of the snout.

Stomach well defined, cœcal.

Pyloric appendages present.

Intestine with two turns.

Anus situated under the throat and forward of the pectorals.

Ovary single. (Placed by the side of the stomach in Amblyopsis and Typhlichthys and behind it in Chologaster.)

Viviparous. (Amblyopsis.)

Testes paired. (Amblyopsis.)

Air bladder with pneumatic duct. (Amblyopsis.)

Liver with the left lobe very large and partially enclosing the stomach.

Amblyopsis DeKay, Fishes of New York, p. 187, 1842.

Eyes rudimentary and imbedded under the skin.

Head with numerous transverse and longitudinal rows of sensitive papillæ provided with nerve branches, many of the nerve branches terminating as free filaments outside the papillæ. Small granulations on the spaces between the papillary ridges. Canals under the skin.

Teeth minute, curved, and arranged in rows on the intermaxillary, inferior maxillary and palatine bones.

Body with a prominent papilla just over the opercular opening, at the base of a small papillary ridge, similar to those on the head. Papillary ridges on sides of body of same character as those on the head, and arranged at nearly equal distances from opercular opening to base of caudal fin.

Pyloric appendages, one on each side.

Ovary situated on the right side of the stomach.

Fins. Ventrals small and placed near the anal fin. Dorsal, 9. Anal, 9. Pectoral, 11. Ventral, 4. Caudal, 24.

Amblyopsis spelæus DeKay. LARGE BLINDFISH.

CRAIGE, Procd. Acad. Nat. Sci. Philad., i, p. 175, 1842. DeKay, Fishes N. Y., p. 187, 1842. WYMAN, Amer. Jour. Sci., xlv, p. 94, 1843; Ann. Mag. Nat. Hist., xii, p. 298., 1843. THOMPSON, Ann. Mag. Nat. Hist., xiii, p. 111, 1844. TELLKAMPF, Müller's Arch. f. Anat., p. 392, 1844; N. Y. Jour. Medicine, v, p. 84, with plate, giving three figs. of the fish; position of internal organs; brain; stomach; air bladder; scale (profile view gives the fish *without* ventral fins, but ventral view shows them), 1845. STORER, Synopsis N. A. Fish, p. 435, 1846. OWEN, Lect. Comp. Anat. Fishes, pp. 175, 202 (fig. of brain), 1846. WYMAN, Procd. Boston Soc. Nat. Hist., iii, p. 349, 1850 "DALTON, N. Y. Medical Times, ii, p. 354, 18—." AGASSIZ, Amer. Jour. Sci. xi. p.128, 1851. WYMAN, Procd. Boston Soc. Nat. Hist., iv, p. 395 (1853), 1854; v, p. 18, 1854; Amer. Jour. Sci., xvii, p. 259, 1854 (with figs. of brain, eye, and otolite). GIRARD, Proc. Nat. Sci. Philad., p. 63, 1859. POEY, Mem. de Cuba, ii, p. 104, Pls. 9, 11 (outlines of fish and of brain), 1858. WOOD, Ill. Nat. Hist., iii, p. 314, figure, 1862. TENNEY, Nat. Hist., p. 344, figure, 1865. GÜNTHER, Cat. Fish Brit. Museum, vii, p. 2, 1868. COPE, Ann. Mag. Nat. Hist., viii, p. 368, 1871. PUTNAM, Amer. Nat.,

vi, p. 6 et seq., with figs., Jan., 1872. WYMAN, Mss. notes and drawings in Putnam, Amer. Nat., vi, p. 16 et seq., 1872. PUTNAM, Amer. Nat., vi. p. 116. Feb.,
1872 (additional note on the young).

PLATE 1 (American Naturalist, Vol. vi, Jan., 1872). FIG. 1. Brain, nerves and
organ of hearing of *Amblyopsis spelæus;* enlarged; *a,* olfactory lobes and nerves;
b, cerebral lobes; *c,* optic lobes; *d,* cerebellum; *e,* organ of hearing, showing the
semicircular canals, with the otolite represented in place by the dotted lines; *f,*
medulla oblongata; *g,* optic nerves and eye specks. FIG. 2. Otolite, enlarged.
FIG. 3. Eye, magnified (natural size one-sixteenth of an inch in length); *a,* optic
nerve; *b,* sclerotic membrane; *c,* layer of colorless cells; *d,* layer of pigment cells
(iris?); *e,* lens. FIG. 4. Lens, enlarged and showing the cells. FIG. 5. Eye, enlarged, showing the muscular bands, *a, a, a, a; b,* the lens pressed out of place; *c,*
the optic nerve. FIG. 6. Top of head, showing canals under the skin, natural
size. The two black dots and lines indicate the eyes and optic nerves in position.
FIG. 7. Top of head, showing the arrangement of the ridges of papillæ, nat. size.
FIG. 8. One of the ridges of papillæ from the head, magnified. FIG. 9. Three of
the papillæ from the ridge, still more magnified, showing the cup-shaped summit
and projecting filament. Fig. 10. A portion of the ridge magnified, and treated
with acid, to show the arrangement of the nervous plexus supplying the papillæ
with nervous filaments from a branch (*a*) of the fifth pair. FIG. 11. Epithelial cells
from the head. FIG. 12. Epithelial cells from the body.
PLATE 2. FIG. 1. Natural size; 1*a,* stomach and pyloric appendages, twice nat.
size; 1*b,* scale, magnified (nat. size represented by the small outline on the left
over the figure); 1*c,* abdominal cavity, showing position of stomach and single
ovary, nat. size.

Head more than half as wide as it is long. Length of head, from
tip of jaw to end of operculum, contained nearly twice in length of
body from operculum to base of caudal fin.

Dorsal and anal fins of equal size, rounded, anal commences under
third ray of dorsal.

Pectorals pointed, reaching to commencement of dorsal.

Ventrals pointed, nearly reaching to commencement of anal.

Caudal broad, long and pointed, membrane, enclosing simple rays
above and below, continuing slightly on the tail.

Scales small, longer than broad, with quadrangular centre and from
8 to 12 concentric lines, which are broken and reduced in number an-
teriorly and crossed by numerous radiating furrows posteriorly.*

Colorless, or nearly so, with transparent fins.

Measurements. Largest specimen, 4·5 inches total length. Smallest
specimen, 1·9 total length.

Geographical distribution. Subterranean streams in Kentucky and
Indiana.

Specimens examined: —

PROF. WYMAN'S COLLECTION.
7 specimens. Half grown and adults. Mammoth Cave.

MUSEUM OF COMPARATIVE ZOOLOGY.
7 specimens. No. 778. Half grown and ♂ ♀ adults. Mammoth Cave.
1 specimen. No. —. Two-thirds grown. Cave near Lost River, Orange Co., Ind.

* The scales described were in every instance taken from the 2d or 3d row under the dorsal fin.

BOSTON SOCIETY OF NATURAL HISTORY.
2 specimens. No. 840. Half grown. Mammoth Cave.

PEABODY ACADEMY OF SCIENCE.
1 specimen. No. 520. Adult ♀. Mammoth Cave. Presented to Essex Institute in 1851 by N. Silsbee.

Other specimens. Dr. Günther mentions six specimens and a skeleton in the British Museum. Mr. Thompson, an adult and newly born young in the collection of the Natural History Society of Belfast. Dr. Steindachner has recently sent an adult and eight young to the Vienna Museum. The first specimen of which we have any record was presented to the Academy of Natural Sciences of Philadelphia; the second is the one described by DeKay and then in the Lyceum of Natural History of New York. Prof. Cope obtained three specimens from the waters of Wyandotte Cave in Indiana. Dr. Tellkampf had several specimens from the Mammoth Cave, and it is probable that specimens exist in nearly all the principal museums and in many private collections, as about all that have been caught in the Mammoth Cave for years have been sold by the guides to visitors.

Habits. But little is known of the habits of the large blindfish. Dr. Tellkampf states that they are solitary; on the slightest motion of the water they dart off a short distance, and that they are mostly found near stones or rocks on the bottom, and seldom come to the surface of the water. Prof. Cope states that if they are not alarmed they come to the surface to feed, swim in full sight, and can then be easily captured if perfect silence is preserved. He also thinks that they are principally surface feeders.

In the stomachs of several that I have opened the only remains found were those of Crayfish. In one specimen, opened by Dr. Wyman, a small fish with well developed eyes was found in the stomach. (See Amer. Nat., vi, p. 13, Pl. 1, fig. 13.)

The eggs are well developed in September, and the young are born about the middle to last of October. The young when born are half an inch or less in length, and are *without* external eyes. (See Amer. Nat., Feb., 1872. The young there mentioned may possibly be those of Typhlichthys.)

Typhlichthys GIRARD, Procd. Acad. Nat. Sci. Philad., p. 63, 1859.

Eyes rudimentary and imbedded under the skin.
Head. The same arrangement of rows of sensitive papillæ as in Amblyopsis, and the spaces between the papillæ with granulations as in that genus. (The subcutaneous canals probably exist, but have not yet been made out.)
Teeth, as in Amblyopsis, on the maxillaries and palatines.
Body with papilla over opercular opening, and with the papillary ridges on the sides as in Amblyopsis.
Pyloric appendages one on each side as in Amblyopsis, but of

slightly different proportion and shape.   (Stomach not so pointed behind as in Amblyopsis.)

Ovary situated on right side of stomach, as in Amblyopsis.   (Eggs fewer in number and proportionately larger than in Amblyopsis.)

Fins. Ventrals *absent.* Dorsal, 7 or 8; Anal, 7 or 8; Pectoral, 12; Caudal 24. (This formula is given after counting several specimens. Girard gives, D. 7; A. 8; P. 11; C. 23.)

It will be noticed that the only characters separating this genus from Amblyopsis are the *absence of ventral fins,* the shape of the stomach and pyloric appendages, and larger eggs in less number.

Typhlichthys subterraneus GIRARD.   SMALL BLINDFISH.

GIRARD, Procd. Acad. Nat. Sci. Philad., p. 63, 1859. GÜNTHER, Cat. Fish Brit. Museum, vii, p. 2, 1868 (as a syn. of Amblyopsis). PUTNAM, Amer. Nat., vi, p. 20 et seq., with figs., Jan., 1872.

PLATE 2 (Amer. Nat., Vol. vi., Jan., 1872). FIG. 3, slightly more than natural size; 3a, stomach and pyloric appendages, twice nat. size; 3b, scale, magnified (nat. size represented by small outline over the figure).

Proportions and general appearance, want of color, arrangement of papillary ridges, position and shape of fins as in Amblyopsis spelæus, with the exception that, owing to the jaws being more obtusely rounded, the head is slightly blunter and broader forward.

Membrane of caudal quite prominent and extending forwards to posterior base of dorsal and anal fins.

Scales broader than long. Large quadrangular centre with from 6 to 8 concentric lines reduced in number and broken up on anterior margin. Posterior portion with numerous radiating furrows.

Measurements. Largest specimen, 1·85 inches in total length. Smallest specimen, 1·45 inches in total length.

Geographical distribution. Subterranean streams in Kentucky, Tennessee and Alabama.

Specimens examined: —

MUSEUM OF COMPARATIVE ZOOLOGY.

7 specimens. No. 780. ♂ ♀. Adults. Mammoth Cave. Collected and presented by Alpheus Hyatt, Sept., 1859.
1 specimen. No. 781. Moulton, Alabama. Presented by Thomas Peters.
1 specimen. No. 782. Lebanon, Tennessee. Presented by J. M. Safford.

Other specimens. Dr. Girard described the species from a specimen in the Smithsonian Institution, taken from a well near Bowling Green, Ky. Dr. Günther mentions a specimen, in the British Museum, from the Mammoth Cave.

Habits. Nothing is known concerning the habits of this fish. It is evidently much rarer at the Mammoth Cave than the large species, to judge from the small number in collections. The fact that Mr. Hyatt obtained seven specimens when he was at the cave in September and did not get any of the other species, may indicate some peculiar location in the waters of the cave where it is more abundant than in other places. The eggs were fully developed in these specimens, but no embryos could be detected. The fish is probably viviparous, and very likely gives birth to its young in October.

Chologaster Agassiz, Amer. Jour. Sci., xvi, p. 135, 1853.

Eyes in normal position and well developed.

Head with small granulations on the surface of the skin. (No papillary ridges.)

Teeth minute, curved and arranged in rows on the intermaxillary and inferior maxillary bones. None on the palatines in the adults. (Of the four specimens examined, the two larger (*C. cornutus*) are without palatine teeth, while the single specimen of *C. Agassizii*, which is evidently a young fish, has a few minute teeth on the palatine bones. In the smallest specimen of *C. cornutus* the mouth is abnormal, the intermaxillaries being reduced to a small central portion and there are consequently no teeth in the upper jaw, but the minute teeth on the palatines are present.*)

(*Body* without opercular papilla and papillary ridges on the sides.)

Pyloric appendages two on each side. Stomach rounded and turned slightly on the side.

Ovary situated principally behind the stomach.

Fins. Ventrals *absent.* Dorsal, 8 or 9. Anal, 8 or 9. Pectoral, 12. Caudal, 28.

This genus principally differs from Amblyopsis and Typhlichthys by the presence of eyes, the absence of papillary ridges on the head and body, by having two pyloric appendages on each side instead of one, and by the posterior position of the ovary. It agrees with Typhlichthys in the absence of the ventrals, and the young further agree by the presence of palatine teeth.

---

* I believe this is one of those interesting cases where one set of organs, or one portion of the animal structure, takes the place of another which from accident is wanting, and that in all probability these palatine teeth, that under normal conditions would be cast off as the fish attained maturity, would have continued to exist in this specimen and answer all the purposes of the intermaxillary teeth. But that in this accidental continuance of these palatine teeth, from the mere mechanical use forced upon them, we have the first stages of the development of a distinct genus, to be characterized by permanent teeth on the palatines, and reduced upper jaw bones, as many of the developmental school would argue, I do not think will bear the test of facts observed.

A not uncommon malformation of fishes consists in the entire or partial absence of the maxillary or intermaxillary bones. I have specially noticed this among our common fresh water trout (Salmo) and marine conner or sea perch (Ctenolabrus) but there have never been recorded allied genera with these characters, while the malformed specimens are hardly numerous enough to give support to the theory that such malformations are hereditary, and it is probable that each case was caused by the non-development of the parts from special cause during embryonic life, or by accident to the individual.

Chologaster cornutus Agassiz.

Agassiz, Amer. Jour. Sci., xvi. p. 135, 1853. Girard, Procd. Acad. Nat. Sci. Philad., p. 63, 1859. Günther, Cat. Fish. Brit. Museum, vii, p. 2, 1868. Putnam, Amer. Nat., vi, p. 21 et seq., with figs. Jan., 1872.

Plate 2 (Amer. Nat., Vol. vi, Jan., 1872). Fig. 2. Natural size. 2a, stomach and pyloric appendages, twice nat. size. 2b, scale magnified (nat. size represented by small outline over the left of the fig). 2c, abdominal cavity showing stomach and single ovary behind the stomach, twice nat. size.

Head more than half as wide as it is long. Length of head, from tip of under jaw to end of operculum contained twice in length of body from operculum, to caudal fin. Width between the eyes equal to distance from eye to tip of under jaw.

Eyes of moderate size, situated just back and over the end of the maxillaries.

Dorsal and anal fins of nearly equal size, slightly rounded. Anal with slightly longer rays and commences under fourth ray of dorsal.

Pectoral fins pointed, reaching to line of commencement of dorsal.

Caudal fin pointed, about equal in length to the head. Membrane above and below extending but slightly on the tail.

Scales very small and deeply imbedded in the skin. Circular with small smooth space forward of the centre. From 15 to 20 concentric rings, cut by a few short radiating furrows on anterior, and longer and more numerous ones on posterior margin.

Intestine is a little longer than in an Amblyopsis of the same size. The two pyloric appendages on the left side are close together and broader than the two on the right side, which are wider apart, longer and more slender than the others.

Color. Yellowish brown, much darker above, lighter on sides, and light yellow on under part and sides of head, belly and under part of tail. Three longitudinal very dark brown lines on each side: the upper commencing near the middle of top of head and following along the back to base of caudal fin; the middle one commencing at the nostril and passing through the eye to upper portion of operculum, thence about in the centre of side to about the centre of base of caudal fin; the lower commences under the pectoral fin and follows the ventral curve of the body to the base of caudal fin. All three lines are darkest and broadest forward, and terminate as a series of nearly confluent dots on the tail. Central rays of the caudal dark brown, outer rays uncolored. Dorsal, anal and pectorals not colored.

Measurements. The three specimens are respectively 1·5, 2, and 2·3 inches in total length.

Geographical distribution. South Carolina.

Specimens examined :—

Museum of Comparative Zoology.
3 specimens. No. 776. Rice Ditches at Waccamaw, S. C. Presented by P. C. J. Weston. 1853. (Orig. of Agassiz.)

Habits. Nothing is known concerning the habits of this species, the only specimens observed being the three mentioned. From the fact of its having a single ovary containing a small number (about 60) of large eggs it is probable that it is viviparous.

Chologaster Agassizii Putnam.

PUTNAM, Amer. Nat., vi, p. 22 et seq., with figs. Jan., 1872.

PLATE 1 (Amer. Nat., Vol. vi, Jan., 1872). FIG. 4. Natural size; 4*a*, stomach and pyloric appendages, twice nat. size; 4*b*, scale magnified (nat. size shown by minute dot over left of the figure).

Head more than half as wide as it is long. Its length is contained three times in the length of the body from the operculum to the base of caudal fin.

Eyes proportionately large and placed over ends of maxillaries.

Dorsal and anal fins broken, but probably of about equal size. Anal fin commences about under fourth ray of dorsal.

Pectoral fins pointed and reaching about half way to the dorsal.

Caudal fin pointed, not quite as long as the head.

Scales very minute, longer than wide, with 4 or 5 concentric lines round a granulated centre. A few radiating furrows cut the concentric lines on the posterior margin.

Pyloric appendages and stomach about the same as in *C. cornutus.*

Color. Uniform light brown, without markings except that the base of the caudal fin is rather darker than rest of fish. Fins uncolored.

Measurements. Total length, 1·4 inches.

Geographical distribution. Subterranean streams in Tennessee.

Specimen examined:—

MUSEUM OF COMPARATIVE ZOOLOGY.

1 specimen. No. 777. From a well in Lebanon, Tenn. Presented by J. M. Safford. Jan., 1851.

This species principally differs from *C. cornutus* by having a longer body and smaller head, by having the eyes proportionately larger, and by its coloration. Nothing is known of its habits except the fact of its subterranean life. The scales of the single specimen known indicate a young fish, and it is probably not over half grown.

———

The four species given in this synopsis are all of the family as yet known, but that others will be discovered and the range of the present known species extended is very probable. The ditches and small streams of the lowlands of our southern coast will undoubtedly be found to be the home of numerous individuals, and perhaps of new species and genera, while the subterranean streams of the central portion of our country most likely contain other species.

www.ingramcontent.com/pod-product-compliance
Lightning Source LLC
Chambersburg PA
CBHW022007190326
41519CB00010B/1419